Andreas Zerfaß

Learning Rates of Electric Vehicles

Anchor Academic
Publishing

Zerfaß, Andreas: Learning Rates of Electric Vehicles, Hamburg, Anchor Academic Publishing 2017

Buch-ISBN: 978-3-96067-177-0
PDF-eBook-ISBN: 978-3-96067-677-5
Druck/Herstellung: Anchor Academic Publishing, Hamburg, 2017

Bibliografische Information der Deutschen Nationalbibliothek:
Die Deutsche Nationalbibliothek verzeichnet diese Publikation in der Deutschen Nationalbibliografie; detaillierte bibliografische Daten sind im Internet über http://dnb.d-nb.de abrufbar.

Bibliographical Information of the German National Library:
The German National Library lists this publication in the German National Bibliography. Detailed bibliographic data can be found at: http://dnb.d-nb.de

All rights reserved. This publication may not be reproduced, stored in a retrieval system or transmitted, in any form or by any means, electronic, mechanical, photocopying, recording or otherwise, without the prior permission of the publishers.

Das Werk einschließlich aller seiner Teile ist urheberrechtlich geschützt. Jede Verwertung außerhalb der Grenzen des Urheberrechtsgesetzes ist ohne Zustimmung des Verlages unzulässig und strafbar. Dies gilt insbesondere für Vervielfältigungen, Übersetzungen, Mikroverfilmungen und die Einspeicherung und Bearbeitung in elektronischen Systemen.

Die Wiedergabe von Gebrauchsnamen, Handelsnamen, Warenbezeichnungen usw. in diesem Werk berechtigt auch ohne besondere Kennzeichnung nicht zu der Annahme, dass solche Namen im Sinne der Warenzeichen- und Markenschutz-Gesetzgebung als frei zu betrachten wären und daher von jedermann benutzt werden dürften.

Die Informationen in diesem Werk wurden mit Sorgfalt erarbeitet. Dennoch können Fehler nicht vollständig ausgeschlossen werden und die Diplomica Verlag GmbH, die Autoren oder Übersetzer übernehmen keine juristische Verantwortung oder irgendeine Haftung für evtl. verbliebene fehlerhafte Angaben und deren Folgen.

Alle Rechte vorbehalten

© Anchor Academic Publishing, Imprint der Diplomica Verlag GmbH
Hermannstal 119k, 22119 Hamburg
http://www.diplomica-verlag.de, Hamburg 2017
Printed in Germany

Table of contents

Table of contents I
List of figures II
List of tables III
List of abbreviations IV
Abstract 1
1 Introduction 3
2 Methods 5
3 Results 15
    3.1 Analysis of vehicle prices 15
    3.2 Experience curves for BEVs and PHEVs 17
    3.3 Time series analysis of real user costs of BEVs, PHEVs, and CVs 19
    3.4 Costs of abating carbon dioxide by BEVs and PHEVs 24
    3.5 Costs of mitigating NO_X and PN emissions by BEVs and PHEVs 29
4 Discussion 33
    4.1 Strengths and limitations of the research 33
    4.2 Discussion of results 35
5 Conclusions and recommendations 38
References 40
Appendix A: Overview of collected data 79
Appendix B: Vehicle price as function of rated engine power and battery capacity 105
Appendix C: Time series of real absolute vehicle price and rated engine power 111
Appendix D: Benchmarking the costs of mitigating NO_X and PN emissions by BEVs and PHEVs 113

List of figures

Figure 1: Scatter plot of the real specific prices of all BEVs, PHEV and CVs 16
Figure 2: Mean specific price of BEVs, PHEVs, and CVs 17
Figure 3: Experience curves for the real specific price of BEVs and PHEVs 18
Figure 4: Experience curve for the real specific price of BEVs expressed per unit of battery capacity 19
Figure 5: Mean real user costs of BEVs, PHEVs and CVs 20
Figure 6: Real user costs of BEVs disaggregated into four principal cost components 21
Figure 7: Real user costs of PHEVs disaggregated into five principal cost components 21
Figure 8: Real user costs of CVs disaggregated into four principal cost components 22
Figure 9: Share of cost components in the real user costs of BEVs in the year 2016 23
Figure 10: Share of cost components in the real user costs of PHEVs in the year 2016 23
Figure 11: Share of cost components in the real user costs of CVs in the year 2016 24
Figure 12: Real costs of mitigating the CO_2 emissions of conventional vehicles by BEVs and PHEVs 27
Figure 13: Mean real costs of mitigating the CO_2 emissions of conventional vehicles by BEVs and PHEVs 29
Figure 14: Real costs of mitigating the NO_X and PN emissions of conventional gasoline and diesel vehicles by BEVs and PHEVs 31
Figure 15: Real absolute price of BEVs as function of rated engine power 106
Figure 16: Real absolute price of PHEVs as function of rated combined power output of electric engine and combustion engine 106
Figure 17: Real absolute price of CVs as function of rated engine power 107
Figure 18: Real absolute price of BEVs as function of battery capacity 108
Figure 19: Real absolute price of PHEVs as function of battery capacity 109
Figure 20: Battery capacity as a function of electric engine power output (BEV) and system power output (PHEV) 110
Figure 21: Mean absolute prices of BEVs, PHEVs and CVs from 2010 to 2016 111
Figure 22: Mean engine power of BEVs, PHEVs, CVs from 2010 to 2016 112
Figure 23: Costs of mitigating NO_X emissions of conventional vehicles and in the manufacturing industry 114
Figure 24: Costs of mitigating PN emissions of conventional vehicles 115

List of tables

Table 1: Overview of assumptions made for calculating the costs of vehicle use and emissions abatement 9

Table 2: Generic assumptions on the real-word NO_X and PN emissions of BEVs, PHEVs, and CVs 10

Table 3: Principle strengths and limitations of this thesis 33

Table 4: Strengths and limitations of the experience curve analysis 34

Table 5: Strengths and limitations of the analysis of user costs 34

Table 6: Strengths and limitation of the analyfiguresis of costs for mitigating CO2 and air pollutant emissions 35

Table 7: Average fuel and electricity prices 79

Table 8: Yearly inflation rates for motor cars in Germany 79

Table 9: Divergence between certified energy consumption and real world energy consumption of CVs 80

Table 10: Divergence between certified energy consumption and real world energy consumption of PHEVs 81

Table 11: Production and cumulated production of electric vehicles worldwide until 2016 ... 81

Table 12: Price data of BEVs and their equivalent CVs 82

Table 13: Price data of PHEVs and their equivalent CVs 87

Table 14: Monthly fixed and maintenance costs of BEVs and their equivalent CVs 91

Table 15: Monthly fixed and maintenance costs of PHEVs and their equivalent CVs 93

Table 16: Energy consumption and CO_2 emissions of BEVs and their conventional CVs 95

Table 17: Energy consumption and CO_2 emissions of PHEVs and their conventional CVs ... 96

Table 18: Real users costs of BEVs and their equivalent CVs in the specific years 97

Table 19: Real user costs of PHEVs and their equivalent CVs in the specific years 102

Table 20: Data sources and assumptions used for estimating the costs of mitigating NO_X emissions of road vehicles and the manufacturing industry 116

Table 21: Data sources and assumptions used for estimating the costs of mitigating PN emissions of road vehicles and the manufacturing industry 119

List of abbreviations

BEV	*Battery Electric Vehicle*
PHEV	*Plug in Hybrid Electric Vehicle*
CV	*Conventional Vehicle*
CO_2	*Carbon Dioxide*
NO_X	*Nitrogen Oxides*
PN	*Particle Number*
ICE	*Internal Combustion Engine*
WTW	*Well-to-Wheel*
LCA	*Life Cycle Assessment*

Abstract

Governments of many countries consider the electrification of individual passenger transport as a suitable strategy to decrease oil dependency and reduce transport-related carbon dioxide (CO_2) and air pollutant emissions. However, battery-electric vehicles (BEVs) and plug-in hybrid-electric vehicles (PHEVs) have been more expensive than their conventional counterparts and suffer from relatively short electric driving ranges, which still hampers the market potential of these vehicles. Despite persisting shortfalls, mechanisms such as technological learning and economics of scale promise to improve the techno-economic performance of BEVs and PHEVs in the short- to mid-term. Here, we seek to obtain insight into the techno-economic prospects of BEVs and PHEVs by: (i) establishing experience curves and (ii) quantifying user costs and the costs of mitigating carbon dioxide and air pollutant emissions in a time-series analysis. Our analysis captures the situation in Germany between 2010 and 2016.

We find that the specific prices [EUR_{2015}/kW] of BEVs and PHEVs decline at a robust rate of 55% (between 2010 and 2016) and 24% (between 2011 and 2016). In the same periods, the specific price and price differentials relative to conventional vehicles decline at learning rates of $23 \pm 2\%$ and $32 \pm 2\%$ (BEVs) and $6 \pm 1\%$ and $37 \pm 2\%$ (PHEVs), respectively. The relatively high learning rates can be explained by decreasing purchasing prices due to high technological learning of traction batteries. The analysis also reveals a correlation between vehicle price and battery capacity for BEVs but not for PHEVs. The vehicle price constitutes the major component of the real user costs, which remain higher for BEVs and PHEVs compared to conventional vehicles. We find user costs to incline in the period of our analysis by (i) 17%, reaching 0.73 ± 0.46 EUR_{2015}/km in 2016 for BEVs and (ii) 39%, reaching in 2016 0.98 ± 0.39 EUR_{2015}/km for PHEVs. The user costs are in the range of those of conventional vehicles (0.69 ± 0.44 EUR_{2015}/km).

The costs of mitigating CO_2 and air pollutant emissions through BEVs and PHEVs have been generally declining, albeit subject to the scenario considered. By 2016, BEVs mitigate CO_2 emissions relative to conventional vehicles at costs of 1520 ± 3325 EUR_{2015}/t (considering certified CO_2 emission levels at the tail pipe), 955 ± 2353 EUR_{2015}/t (considering actual on-road CO_2 emissions at the tail pipe), 962 ± 7837 EUR_{2015}/t (considering CO_2 emissions along the entire well-to-wheel (WTW) chain), and 1789 ± 21412 EUR_{2015}/t (considering CO_2 emissions along the WTW chain and those resulting from battery manufacturing). PEHVs emit more CO_2 than their conventional counterparts when considering a WTW scenario including emissions from battery manufacturing. In 2016, BEVs mitigate nitrogen oxides (NO$_X$) at costs of 2486 ± 5995 EUR$_{2015}$/kg NO$_X$ and 459 ± 709 EUR$_{2015}$/kg NO$_X$, depending on whether the emission factors of gasoline or diesel vehicles are considered. Particle number emissions are mitigated at costs of 0.004 ± 0.01 EUR$_{2015}$/10^{11} particles and 0.23 ± 0.36 EUR$_{2015}$//10^{11} particles, assuming emission factors for gasoline and diesel vehicles, respectively. The costs of mitigating air pollutant emissions through PHEVs are generally higher and depend on the mode of vehicle operation and on whether the PHEV is propelled by a gasoline or diesel engine. We do not find particles emission savings of diesel-propelled PHEVs over CVs.

We regard our findings as robust but acknowledge that the costs of emission mitigation is subject to uncertainty as vehicle use and electricity generation can vary by large margins within and between countries. Furthermore, we might slightly overestimate the user costs due to the assumption of a relatively short lifespan compared to common lifecycle-assessment practices. The robust decrease in price differentials coupled with the financial incentives provided in many European countries together suggest that electric vehicles may soon be price and cost competitive even without drastic further price drops. It remains therefore important to focus on other obstacles for the deployment of BEVs and PHEVs, like the implementation of a widespread (quick) charging infrastructure, the extension of driving range, and the durability of traction batteries. Policy makers should be aware of rebound effects through which cost and emission savings may be absorbed by the purchase of more powerful luxury cars.

Analyses for a complete lifecycle scenario for the vehicles themselves could be made in future researches. In addition, an analysis excluding all taxes, charges, subventions and other governance could be made in future researches but the implementation could suffer from to many hidden influences such as subventions along the WTW chain that is almost impossible to exclude.

1 Introduction

Concerns about air pollution, anthropogenic greenhouse gas emissions, and the dependency on non-renewable fossil fuels have been facilitating the electrification of passenger road transport in many regions of the world (e.g., EC, 2011; Masiero et al., 2016; TWH, 2016). The energy use for transport has doubled over the past three decades (Dulac, 2012), accounting in 2012 for about a quarter of the final energy consumption worldwide (IEA, 2016a). Light-duty passenger vehicles consume the bulk of this energy, nearly accounting for the combined energy use of all other modes of road, rail, water, and air transport together (IEA, 2016a). In consequence of the increasing energy use, the carbon emissions of the transport sector have been increasing over the past decades and are expected to do so by another 80% between 2007 and 2030 (Kahn et al., 2007). The contribution of vehicle emissions to air pollution (e.g., Huang et al., 2011; EEA, 2016; Degraeuwe et al., 2016) remains a concern, as 92% of the global population lives in places where pollutant concentrations exceed the air quality standards of the World Health Organization (WHO, 2016).

Parts of these problems could be addressed by electric vehicles that do not exhibit tail-pipe emissions and that could be driven with electricity generated from renewable sources. Yet, electric vehicles are still comparatively expensive and suffer from short drive ranges and an insufficient recharging infrastructure. Policy makers around the world began addressing these shortcomings through dedicated support policies (IEA, 2016b), complemented by concrete goals for the market penetration of electric vehicles (IEA, 2016b). The German government, for instance, targets a stock of 1 million electric vehicles by 2020 (BR, 2016), to be achieved through subsidies and partial exemptions of electric vehicles from road taxation (BMWi, 2017). China and the USA likewise have implemented ambitious targets of 5 million (SC, 2012) and 1.2 million (IA-HEV, 2015) electric vehicles, respectively on the road by 2020. Until 2016, the global fleet of electric cars has, in fact, been growing exponentially from a few thousand vehicles in 2009 to 1.3 million. Yet, its 0.1% share in the fleet of nearly 1 billion passenger cars is still negligible (IEA, 2016b).

We would expect that the share of electric vehicles in both the market and in-use fleet can substantially increase, if their price decreases. Knowing about economies of scale and the potentials for technological learning, one would expect, that manufacturing costs, and hence vehicle prices, have been declining in the past and will continue to do so in the future. Following this hypothesis, user costs and the costs of mitigating carbon dioxide (CO_2) and air pollutant emissions may have been declining, thus improving greatly the economic efficiency of electric vehicles in addressing environmental shortfalls of road transport. However, the rate of price decline and the trends in user costs as well as the costs to mitigate CO_2 and air pollutant emissions through electric vehicles have not yet been investigated.

Here, we address this knowledge gap in a step-wise approach. First, we analyze the prospects for a future decline in the price of electric vehicles by means of the experience curve approach. Experience curves quantify technological learning by modelling the production costs of a technology as a power-law function of cumulative production. Experience curves were previously established for hybrid vehicles (Weiss et al., 2012), electric two-wheelers (Weiss et al., 2015), and vehicle batteries (Nykvist and Nilsson, 2015). Here, we establish experience curves capturing the price dynamics of battery-electric (BEVs) and plug-in hybrid-electric vehicles (PHEVs) for the period between 2010 and 2016.

Second, we quantify for BEVs, PHEVs, and conventional vehicles (CVs) the cost of ownership [EUR_{2015}/km] in a time-series analysis capturing the period between 2010 and 2016.

Third, we quantify for BEVs, PHEVs, and CVs the energy use [kWh/km], CO_2 emissions [gCO_2/km], air pollutant emissions [mg NO_X/km; particle number/km] for various scenarios, e.g., considering certified emission levels, actual on-road emissions, the WTW energy chain, and an extended WTW scenario that also includes the CO_2 emissions from battery manufacturing (Moro and Helmers, 2015). Based on this information, we establish time-series analyses of the costs of mitigating CO_2 and pollutant emissions by BEVs and PHEVs in the period between 2010 and 2016. The third part of our analysis can indeed reveal new insights into the cost-effectiveness of BEVs and PHEVs in mitigating CO_2 and air pollutant emissions.

The thesis continues with a description of our research method in Section 2. We present results in Section 3 and provide a discussion of the strengths and limitations of our research in Section 4.1 and discuss the results in Section 4.2. The article finishes with conclusions and recommendations for scientists and policy makers in Section 5.

2 Methods

2.1 Definitions

Throughout this thesis, we use the terms 'electric vehicle' and 'battery-electric vehicle' (BEV) synonymously for passenger cars that are exclusively propelled by one or multiple electric engines, drawing their propulsion energy solely from an electric energy storage system such as a battery.

Plug-in hybrid electric vehicles (PHEVs) are defined as passenger cars that: (i) are equipped with an internal combustion engine (ICE) and one or multiple electric engines, (ii) draw their propulsion energy from combustible fuels and/or electricity, and (iii) can be charged from an external electricity source (UNECE, 2015). We do not distinguish between *parallel* PHEVs in which the internal combustion engine and the electric engine are both connected to a transmission and can thus propel the vehicle in parallel and *series* PHEVs (also referred to as range-extender vehicles) in which the electric engine propels the vehicle, whereas the internal combustion engine functions as an electricity generator to charge the battery. Our choice is justified by the limited number of PHEV models offered on the market in each individual year of our analysis. We acknowledge that not distinguishing between *parallel* and *series* PHEVs introduces uncertainty into our analysis because parallel PHEVs equipped with a small, thus less costly, battery and a full-size internal combustion engine are lumped together with *series* PHEVs equipped with a comparatively large, thus costly, battery and a rather small internal combustion engine. However, the database contains only around 8.5% (5 out of 59) vehicles with range extender, meaning that PHEVs take the major part in the analysis.

Throughout this thesis, we refer to 'conventional vehicles' (CVs) as passenger cars propelled exclusively by an internal combustion engine that draws its energy from combustible fuels such as gasoline or diesel.

2.2 Data collection

In our analysis, we include BEVs, PHEVs, and comparable CVs sold in Germany in the period between 2010 and 2016. Having used an extended web search we start out by identifying all models of BEVs, PHEVs, and their respective conventional counterparts that are produced in series and offered for sale on the German market in each individual year of our analysis. Regarding BEVs, we only include vehicles for which the given price includes the traction battery. For example, the manufacturer Renault sells its BEVs without traction battery and instead charges a monthly leasing fee for the battery. Vehicles with lease battery are excluded from our data collection because their prices are not comparable to the battery-including prices stated for other BEVs. Moreover, we exclude from our analysis BEVs and PHEVs that are: (i) produced in limited series (i.e., less than 1,000 vehicles per year) and (ii) intended for racing rather than passenger transport (e.g., Porsche 918 Spyder; McLaren P1).

For each vehicle included in our analysis, we collect data on the sales price[1] [EUR], maintenance costs [EUR], fixed costs related to insurance and vehicle registration in Germany [EUR], engine power [kW], if applicable, the capacity of the traction battery [kWh], certified distance specific energy consumption [kWh/km; l/100km], CO_2 emissions [gCO_2/100km] and the certified emissions standard as published by the miscellaneous manufacturers on their web pages, in their product brochures, or as published by third parties like newspaper articles or web pages (see Tables 16 and 17 in Appendix A). If there is no price for a certain vehicle model in a given year available, the price of the previous year has been assumed. This assumption can be made because manufacturers tend to announce price changes on their web pages; relevant information becomes also available via third parties like newspapers. Equivalent conventional vehicles are chosen, as far as feasible, to match the year of production, vehicle category, manufacturer name, model, type, size, as well as engine power of BEV and PHEV models in each year. We generally chose CVs with a manual transmission. This choice takes into account that not all consumers who purchase a BEV or PHEV may have chosen otherwise for a CV with automatic transmission but instead might have preferred a less costly, but otherwise equivalent, CV with a manual transmission. Following this argument, we consider the calculated price and cost differentials between BEVs/PHEVs and CVs to reflect somewhat a 'worst-case scenario'. Moreover, electric cars don't have a transmission that could increase energy consumption; by contrast automatic transmissions on conventional cars might unduly increase the fuel consumption.

Information on the value added tax in Germany is obtained from Statista (2017); the yearly inflation rate in Germany is obtained by Eurostat (2017; see Table 8 in Appendix A). More precisely, the specific inflation rates for motor cars are taken into consideration.

For our experience curve analysis, we furthermore assume a yearly global production of electric cars based on the number of yearly new registrations obtained from ZSW (2016; see Table 11

---

[1] The purchasing prices in our analysis do not include the subsidy buyers of BEVs and PHEVs obtain in Germany since July 2016 (The Federal Government, 2017) as this would have distorted the price progression.

in Appendix A) with BEVs and PHEVs ascertained together as they both contain a traction battery being accountable for large portion of the production costs.

For the calculation of use-phase cost and the costs of mitigating CO_2 and air pollutant emissions, we collect data on:

- the yearly costs of vehicle maintenance and fix costs (ADAC, 2017; see Table 14 and Table 15 in Appendix A);
- the generic prices of fuel (diesel and gasoline) and electricity that is kept constant for all years of our analysis (see Table 1);
- the life time and yearly mileage of BEVs, PHEVs, and CVs (see Table 1);
- the real-world electricity consumption of BEVs, as well as the real-word CO_2 emissions at the tailpipe of PHEVs and CVs (see Table 1);
- the CO_2 and NO_X emissions caused by the electricity production (see Table 1);
- the CO_2 emissions of battery production (see Table 1);
- the real-word NO_X and particle number emissions of PHEVs and CVs (Table 2).

The fix costs provided by ADAC contain an indemnity insurance with a contribution rate of 50% for the region group 6, regarding the car classification as well as a fully comprehensive cover with 500 EUR co-payment with a contribution rate of 50% for the region group 6, considering standard charges without any additional discounts. The fix costs also contain vehicle tax, however, tax exemptions for electric vehicles are considered. Further, a fixed rate of 200 € per year for, e.g., parking prices, general inspections, exhaust analyses, and minor accessories is added.

The maintenance costs contain costs for oil changes, inspections, wear and tear repairs, tires and for vehicles older than three years and additional fixed rate staggered by car classifications.

The life time of all vehicles contained in this analysis is set to 6 years because the ADAC (2017) data about the maintenance costs, which would increase with vehicle age, are based on a similar life time. The assumption of a relatively short life time of 6 years may also be justified in view of the uncertain durability of electric batteries and the likely need to replace batteries if longer life times are assumed (for a discussion, see Helmers and Weiss, 2017). In addition, the German ministry of finance also sets a period of six years for the depreciation of commercially used cars. We acknowledge that vehicles may have a substantially longer life time than 6 years; the assumption of a comparatively short lifespan leads to a higher influence of the vehicle purchasing price on the user costs in comparison to the other variables such as fuel costs.

We abstain from considering the results of road-side remote-sensing measurements. Remote sensing has been frequently applied to capture the emissions behavior of a large number of vehicles under real-world conditions (e.g., Carslaw et al., 2011; Chen and Borken-Kleefeld, 2014, 2016) Yet, measurements (i) have been focused on pre-Euro 6 vehicles, (ii) only capture a snap shot of the emissions performance of each vehicle at a specific location and under specific driving conditions, (iii) and require making assumptions about the instantaneous fuel use to be used for determining distance-specific emission factors [g/km] (Franco et al., 2013). The results of remote-sensing measurements may not in all instances accurately capture the average emissions behavior of vehicles. We differentiate between gasoline and diesel vehicles as well as vehicles certified according to the Euro 5 and 6 emission limits as NO_X and particle number emissions differ between these vehicle categories. The Euro 5 standard is assumed to apply to all vehicles until 2015; the Euro 6 standard applies to all vehicles sold in the years 2015 and 2016 (Euro 6 has been mandatory for new vehicles since 2015 (EC, 2007)). It is noteworthy that the proportionality between energy consumption and NO_X as well as particle number emissions is not as obvious as the proportionality between fuel consumption and CO_2 emissions because former emissions are a function of various emissions control technologies.

Table 1: Overview of assumptions made for calculating the costs of vehicle use and emissions abatement

	BEVs	Source	PHEVs	Source	CVs	Source
Lifetime [a]	6	BMF (2017)	6	BMF (2017)	6	BMF (2017)
Yearly mileage [km]	14259	KBA (2015)	14259	KBA (2015)	14259	KBA (2015)
Electricity price [EUR/kWh]	0.27	BDEW (2017)	0.27	BDEW (2017)	-	-
Fuel price [EUR$_{2015}$/l]	-	-	1.31 (diesel) 1.49 (gasoline)	SB (2017)	1.31 (diesel) 1.49 (gasoline)	SB (2017)
Carbon intensity of the electricity mix in Germany (including losses from plant to plug and own consumption in power plants [g CO_2 eq./kWh]	707	Helmers et al. (2017)	-	-	-	-
Average carbon losses along the fuel production and distribution chain [% of certification value]	-	-	18	Cames et al. (2013)	18	Cames et al. (2013)
Correction factor for real-world electricity and fuel consumption [% of certification value]	30	Zerfass (2015)	217.6	based on averaged ICCT (2016b) data for PHEVs, see Table 9	See Table 9	See Table 9
Powerplant NO$_X$ Emissions (g/kwh)	0.44	Helmers, E. (2010)	0.44	Helmers, E. (2010)	-	-
Emissions impact of battery production [kg CO_2eq./kWh]	168	Helmers et al. (2015)	168	Moro et al. (2015)	-	-

The NO$_X$ and particle number emission factors listed in Table 2 represent generic factors that account for the actual on-road tail-pipe emissions of PHEVs and CVs. The applied NO$_X$ emission factors are based on the European Environmental Agency's air pollutant emission inventory guide book (EEA, 2016) that is used by the COPERT model to calculate air pollutant emissions from road transport. The emission factors are within the range of values identified during on-road testing with Portable Emission Measurement Systems (e.g., Weiss et al., 2012; Yang et al., 2015). Measurements of PN emissions are still scarce. The PN emission factors applied here are based on a limited number of tests conducted on the chassis dynamometer and on the road by Giechaskiel et al. (2015).

Table 2: Generic assumptions on the real-word NO$_X$ and PN emissions of BEVs, PHEVs, and CVs

Pollutant	NO$_X$ [mg/km]	PN [#/km]
Source	EEA (2016)	Giechaskiel et al. (2015)
BEVs	0	0
PHEVs - Gasoline (Euro 5)	13	8·10^{11} a
PHEVs - Gasoline (Euro 6b)	13	3·10^{12} b
PHEVs - Diesel (Euro 5)	490 f	8·10^{11} g
PHEVs - Diesel (Euro 6b)	490 f	8·10^{11} g
CVs – Diesel (Euro 5)	610	4·10^{11} c
CVs – Diesel (Euro 6b)	500	4·10^{11} c
CVs – Gasoline (Euro 5)	60	1·10^{12} d
CVs – Gasoline (Euro 6b)	60	4·10^{12} e

[a] estimate based on the PN emissions of one car equipped with a port-fuel-injection spark-ignition engine and the assumption that 20% of the distance are driven by PHEVs electrically
[b] estimate based on the midpoint of PN emissions observed for seven vehicles with gasoline direct injection engines and the assumption that 20% of the distance are driven by PHEVs electrically
[c] estimate based on the mean PN emissions of two diesel cars equipped with a particle filter
[d] estimate based on the PN emissions of one car equipped with a port-fuel-injection spark-ignition engine
[e] estimate based on the midpoint of PN emissions observed for seven vehicles with gasoline direct injection engines
[f] estimate based on the two plug-in hybrid diesel vehicles tested by Franco et al. (2016)
[g] conservative estimate based on expert judgment and two plug-in hybrid diesel vehicles tested by Hammer et al. (2015)

2.3 Data analysis

We begin our analysis by calculating for each vehicle the real price $P_{i,t}$ [EUR$_{2015}$] deflated or inflated to the base year 2015 as:

$$P_{i,t} = \frac{p_{i,t}}{(1+r_t) \times k_t} \tag{1}$$

where $p_{i,t}$ represents the nominal price of vehicle i in year t [EUR], r_t represents the value added tax rate in year t, and k_t represents the year-specific deflator calculated based on the yearly inflation rates in Germany (see Table 8 in Appendix A).

We abstain from analyzing absolute vehicle prices as these can vary greatly depending on the class and size of vehicles. By calculating real specific vehicles prices, we normalize the absolute vehicle prices with the rated engine power of each vehicle. Appendix B demonstrates a linear relationship between the two parameters, suggesting the latter is a suitable functional unit for our price analyses.

Based on the real vehicle price, we calculate for all vehicles the specific price per kilowatt rated engine power [EUR$_{2015}$/kW] and for all BEVs the specific price per kilowatt-hour energy storage capacity of the traction battery [EUR$_{2015}$/kWh]. We abstain from calculating the specific price per kilowatt-hour energy storage capacity for PHEVs as the battery capacity of these vehicles varies over a wide range depending on the power of the complementary internal combustion engine.

With the specific price [EUR$_{2015}$/kW], we calculate the specific price differential ΔP between BEVs and PHEVs and their equivalent conventional vehicles as:

$$\Delta P = P_E - P_C \tag{2}$$

where P_E represents the specific price of BEVs and PHEVs [EUR$_{2015}$/kW] and P_C represents the specific price of conventional vehicles [EUR$_{2015}$/kW].

Based on the calculated values, we explore price trends through a simple time-series analysis and we establish experience curves. For the experience curve analysis, we determine in a first step for each year the mean and standard deviation of specific prices and price differentials of BEVs and PHEVs.

In a second step, we establish separate experience curves for the specific prices and price differential by plotting the yearly mean of each price parameter $P_t(x_t)$ for BEVs and PHEVs as a power-law function of cumulative vehicle production:

$$P_t(x_t) = P_0(x_0) \left(\frac{x_t}{x_0}\right)^b \qquad (3)$$

where $P_0(x_0)$ represents the specific price or price differential of BEVs or PHEVs in the base year of tour analysis; x_0 and x_t represent the cumulative production in the base year and in year t of our analysis, respectively; b represents the experience index. Applying the logarithmic function to Equation 3 yields a linear equation with b as slope parameter and $logP_0(x_0)$ as the intercept of the y-axis. We calculate the learning rate LR [%], that is the rate at which specific prices and price differentials of BEVs and PHEVs decline with each doubling of cumulative production, as:

$$LR = (1 - 2^b) \cdot 100\% \qquad (4)$$

The error interval of LR uniformly represents the standard error of the slope parameter b. We complement the data analysis described so far with a scoping study of the literature on techno-economic information that can explain the results of our analysis.

Next to our analysis of the specific prices and price differentials, we calculate the distance-specific user costs $C_{i,t}$ [EUR$_{2015}$/km] of each BEV, PHEV, and CV i sold in year t as:

$$C_{i,t} = \frac{P_{i,t} + (C_{M,i} + M_i \cdot F_i \cdot C_F \cdot 0.01) \cdot L_i}{M_i \cdot L_i} \qquad (5)$$

where $P_{i,t}$ represents the real vehicle price [EUR$_{2015}$], $C_{M,i}$ the yearly maintenance costs [EUR$_{2015}$/a], M_i the yearly driving distance [km/a], F_i the distance-specific electricity or fuel use [l/100 km; kWh/100km], C_F the price of fuel or electricity [EUR$_{2015}$/l; EUR$_{2015}$/kWh], and L_i the lifetime [a] of each respective vehicle i. We assume constant and averaged values for fuel and electricity prices, which we calculate based on statistical data (SB, 2017), see Table 7 in Appendix A.

Based on the distance-specific user costs and the distance-specific CO_2 and pollutant emissions of vehicles, we calculate the emission abatement costs $EC_{BEV/PHEV}$ [EUR$_{2015}$/100 g CO_2; EUR$_{2015}$/100 mg NO$_X$; EUR$_{2015}$/10^{11} particles] of each BEV and PHEV as:

$$CE_{i,t} = \frac{C_{i,t}(BEV - PHEV) - C_{i,t}(CV)}{E_{i,t}(CV) - E_{i,t}(BEV - PHEV)}$$

(6)

where $C_{BEV/PHEV}$ and C_{CV} represent the distance-specific costs [EUR$_{2015}$/km] of BEVs/PHEVs and CVs, respectively; $E_{BEV/PHEV}$ and E_{CV} represent the CO$_2$ or pollutant emissions [g CO$_2$/km; 100 mg/km; 10^{11} particles/km] of BEVs/PHEVs and CVs, respectively.

With the calculated specific prices, user costs, and the costs of mitigating CO$_2$ and air pollutant emissions, we calculate yearly means and standard deviations and conduct simple time-series analyses to assess whether BEVs and PHEVs have become economically more viable relative to conventional vehicles in the period between 2010 and 2016. We abstain from applying the experience curve approach to the user costs and the costs of mitigating CO$_2$ and air pollutant emissions because these parameters depend on energy prices and the electricity and fuel use of vehicles, i.e., factors that are not directly subject to technological learning but rather related to the size of the vehicle and the power rating of the engine.

For our time-series analyses of the costs to mitigate CO$_2$ emissions, we account for the observation (i) that real-world emission levels often exceed the values declared by the vehicle manufacturers and (ii) that emissions along the energy supply chain as well as from battery manufacturing can contribute substantially to the overall life-cycle CO$_2$ emissions of vehicles. We therefore consider four scenarios: (i) the distance-specific CO$_2$ emissions of vehicles as certified during standard type-approval testing, (ii) the distance-specific CO$_2$ emissions of vehicles under real-word driving conditions based on correction factors proposed by ICCT (2016b), (iii) the distance-specific CO$_2$ emissions along the entire well-to-wheel (WTW) electricity and fuel supply chain for BEVs, PHEVs, and CVs (see Table 1), and (iv) a hybrid WTW scenario as proposed by Moro and Helmers (2015) that accounts next to the CO$_2$ emissions from the electricity and fuel supply chain also for the CO$_2$ emissions from battery manufacturing. We calculate the WTW CO$_2$ emissions for BEVs and PHEVs on the basis of the carbon intensity of the average electricity mix in Germany, including the own consumption of the power plants (Table 1). We assume a WTW correction factor of 18 %, see Table 1. For the hybrid WTW scenario we assume battery manufacturing cases emissions of 168,000 g CO$_2$eq/kWh (Helmers et al., 2015; Table 1). The correction factors are shown in Table 1 and Table 9.

For our time-series analyses of the costs to mitigate air pollution through BEVs and PHEVs, we focus on NO$_X$ and PN as these pollutants represent major concerns for public health (e.g.,

EEA, 2016; WHO, 2016). We generally limit our analysis to tail-pipe emissions but include in one scenario also the NO_X emissions from electricity generation in Germany that cause indirect emissions of BEVs and PHEVs (0.44 g NO_X/kWh based on Helmers (2010); see Table 1).

3 Results

First, we present a simple time-series analysis of the real specific prices of BEVs, PHEVs, and CVs (Section 3.1). Second, we establish *ex-post* experience curves and quantify learning rates for the specific price and price differentials of BEVs and PHEVs (Section 3.2). In the third part, we present time-series analyses of the real user costs per kilometer as well as the real difference in users costs between BEVs and PHEVs and their conventional counterparts (Section 3.3). Fourth, we present the costs of mitigating CO_2 emissions, thereby considering four scenarios, i.e., certified tail-pipe emissions, actual real-world tailpipe emissions on the road, WTW emissions, and a scenario capturing WTW emissions and the emissions from battery manufacturing (Section 3.4). Finally, we present the costs of mitigating NO_X and particle emissions, containing scenarios (i) that capture the tail-pipe emissions of gasoline and diesel vehicles (accounting for Euro 5 and 6 certified vehicles) and (ii) that capture tail-pipe emissions and the NO_X emissions from electricity generation (Section 3.5).

3.1 Analysis of vehicle prices

The specific prices of vehicles scatter over a relatively wide range around the mean for each vehicle category and year, reflecting the diversity of vehicles available on the market (Figure 1). Two sets of outliers are apparent. First, the prices of the BEV Aixam Mega E-City, sold in 2011 as well as in 2015 and 2016, are substantially higher than the prices of the other BEVs. The vehicle is a small, light-weight BEV with only around 5 kW power output, produced as small scale, what taken together explains the very high specific price that we exclude from the following price analyses. Second, the prices of the equivalent CVs Ligier and Aixam Miniauto and Aixam City, which also have only around 5 kW output power and are also produced at small scale, only are equally excluded from the further price analyses.

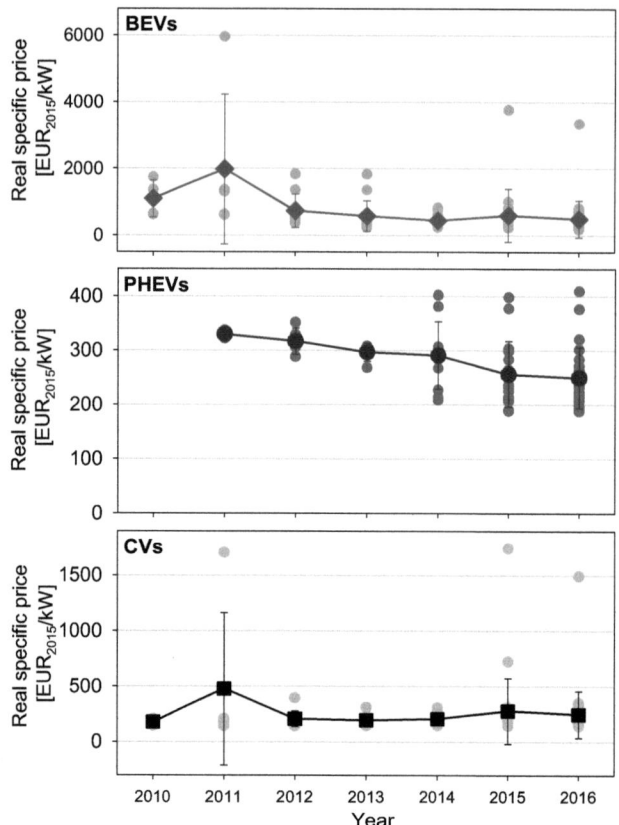

Figure 1: Scatter plot of the real specific prices of all BEVs (top), PHEV (center) and CVs (bottom); bold dots depict the mean price in earch year; error bars represent the standard deviation of price data

The real specific price of BEVs and PHEVs shows a robust decline since 2010 and 2011, respectively. By contrast, the price of equivalent CVs appears to have slightly increased in the same time period (Figure 1). The mean real specific price for BEVs has decreased by around 55% from around 1090 EUR$_{2015}$/kW in 2010 to around 486 EUR$_{2015}$/kW in 2016; the mean real specific price for PHEVs has decreased by around 24% from around 330 EUR$_{2015}$/kW in 2011 to 250 EUR$_{2015}$/kW in 2016, whereas the mean real specific price for CVs has increased by around 39% between 2010 (178 EUR$_{2015}$/kW) and 2016 (248 EUR$_{2015}$/kW; Figure 2). These observations suggest (i) a stronger price decrease for BEVs than for PHEVs, with the former starting at a much higher price level and (ii) a robust decline of the real specific price difference between BEVs and PHEVs and their equivalent CVs. Regarding the development in the period of 2014-2016, the parity of prices of BEVs / PHEVs and CVs will be reached in 2027 / 2019, however, this forecast contains high uncertainty. The rapid price decline for BEVs in the period of 2010-2014 suggests substantial technological learning, which we quantify below.

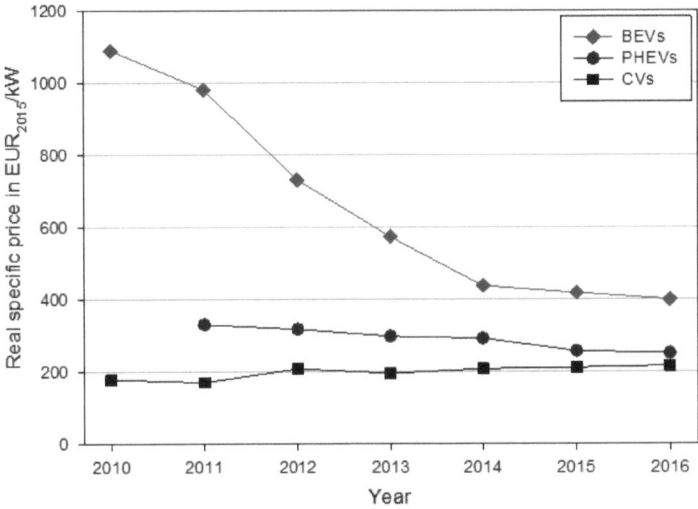

Figure 2: Mean specific price of BEVs, PHEVs, and CVs for all vehicles sold in the specific year; the price outliers identified in Figure 1 are expluded here

3.2 Experience curves for BEVs and PHEVs

In line with the findings above, the experience curve analysis yields a relatively high learning rate of 23 ± 2 % for the specific price of BEVs. By contrast, the specific price of PHEVs only declines at a learning rate of 6 ± 1 %. The high coefficients of determination, i.e., R^2=0.97 and R^2=0.95, respectively suggest a robust experience-curve relationship between specific price and cumulative vehicle production (Figure 3a). We obtain higher learning rates of 32 ± 2% for BEVs and 37 ± 2% for PHEVs when plotting the real specific price differential between BEVs and PHEVs and their equivalent CVs (Figure 3b). The real specific price differential between BEVs and their equivalent CVs has decreased from 920 ± 543 EUR_{2015}/kW in 2010 when around 59,000 BEVs had been produced globally to only 214 ± 237 EUR_{2015}/kW in 2016 when 1.35 million BEVs had been produced. In the same period of time, the real specific price difference between PHEVs and CVs decreased from 182 ± 11 EUR_{2015}/kW in 2011 to only 20 ± 39 EUR_{2015}/kW in 2016, which is de facto parity. We obtain a learning rate of 16 ± 2% when considering the real price of BEVs in relation to the battery capacity [kWh] instead of the maximum rates engine power [kW] (Figure 4). The price per kilowatt-hour battery capacity decreased by 51% from around 2,537 EUR_{2015}/kWh in 2010 to around 1,242 EUR_{2015}/kWh in 2016.

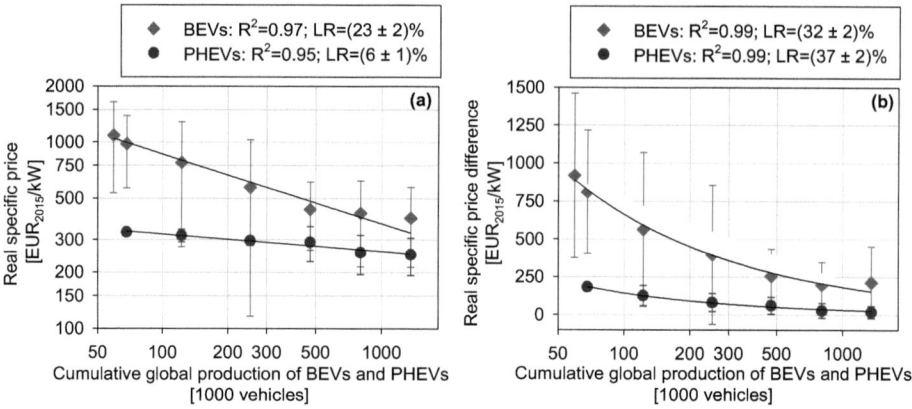

Figure 3: Experience curves for the real specific price of BEVs and PHEVs (a) and for the real specific price difference between BEVs and PHEVs, repsectively and their respective equivalent CVs (b); error bars represent the standard deviation of price data

The learning rates scatter below and above the learning rates of 18 ± 9% (mean ± 95% confidence interval) identified by Weiss et al. (2010) for energy-demand technologies. The learning rates for the specific price of BEVs are considerably higher than:

- the 9% and 12% identified by Safari (2017) for the specific price and the costs of powertrain electrification[2] [USD_{2016}/kWh] of BEVs;

- the 21.6% (Maycock, 2014), 15% (Hoffman, 2014), 17% (Nagelhout and Ros, 2009) and 6-9% (Nykvist and Nilsson, 2015) identified for the specific costs of electric batteries;

- the 8 ± 1% identified by Weiss et al. (2015) for the specific price of e-bikes.

The learning rates for the specific price and price differential of PHEVs are in line with the 7 ± 2% and 23 ± 5% (mean ± 95% confidence interval), respectively identified by Weiss et al. (2012) for non-plug in hybrid vehicles. The high learning rates of BEVs compared to those of PHEVs and non-plugin hybrid vehicles (the latter identified by Weiss et al., 2012) suggest that the decline in the specific price and price differentials of hybrid vehicles may not be a reliable proxy of the price dynamics of BEVs.

---

[2] The costs of the electrification of the powertrain exclude the costs, and thus learning effects, related to the battery (see Safari, 2017).

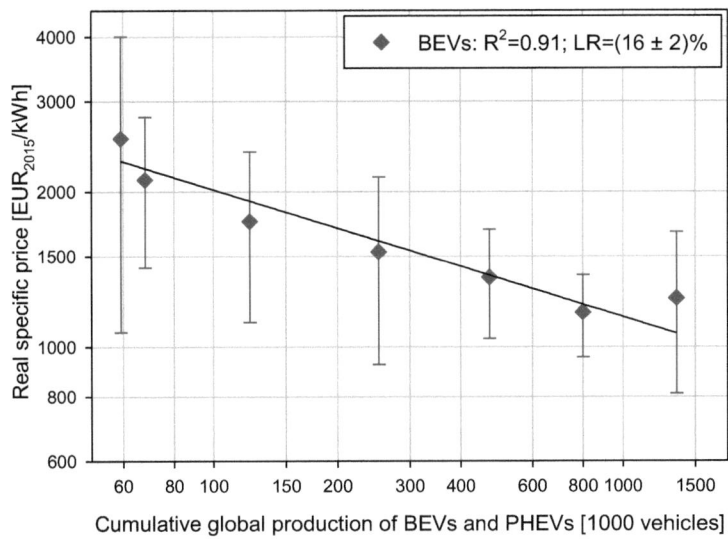

Figure 4: Experience curve for the real specific price of BEVs expressed per unit of battery capacity; error bars represent the standard deviation of price data

3.3 Time series analysis of real user costs of BEVs, PHEVs, and CVs

The real user costs do not follow the trend in specific vehicle prices but tend to remain relatively constant (BEVs) or show an increase (PHEVs and CVs) in the time periods analyzed. This observation suggests that increasing vehicle size and engine power, and thus consumption of electricity and fuel have been compensating the decline in specific vehicle prices (see Appendix C). For BEVs, user costs increased slightly (0.63 ± 0.18 EUR_{2015}/km in 2010 versus 0.73 ± 0.46 EUR_{2015}/km in 2016; Figure 5a). For PHEV, user costs developed from 0.71 ± 0 EUR_{2015}/km in 2011 to 0.98 ± 0.39 EUR_{2015}/km in 2016, which is an increase of 39% (Figure 5a). For CVs the results show a 137% increase of real user costs, from 0.29 ± 0.04 EUR_{2015}/km in 2010 to 0.69 ± 0.46 EUR_{2015}/km. We explain the stagnating or even increasing user costs with increasing vehicle size, engine power, and subsequently distance-specific fuel and electricity use (see Appendix C). The user costs identified here are in line with the findings of Hagman et al. (2016), who reported costs in the rage of 0.42 to 0.47 EUR/km for BEVs.

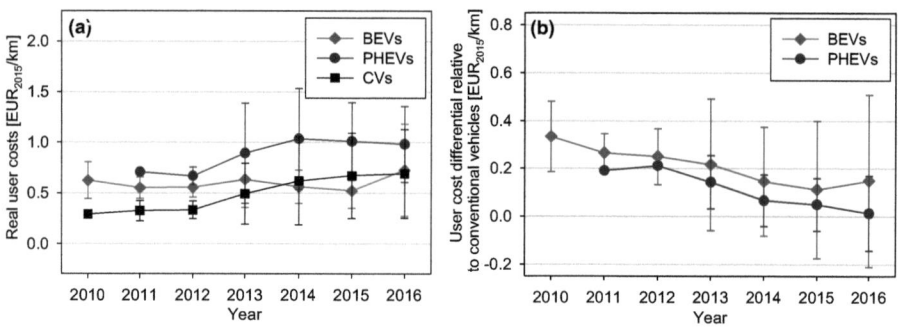

Figure 5: Mean real user costs of BEVs, PHEVs and CVs (a) and mean difference in user costs between BEVs and PHEVs and equivalent CVs (b); error bars represent the standard deviation of cost data

In line with this trend, the additional user costs of BEVs and PHEVs relative to equivalent CVs have decreased in the time period of our analysis (Figure 5b). The additional user costs of BEVs have decreased from 0.33 ± 0.15 EUR$_{2015}$/km in 2010 to 0.15 ± 0.41 EUR$_{2015}$/km in 2016, which is a decrease of 55%. For PHEVs, we observe a decrease of 93%, i.e., from 0.19 ± 0 EUR$_{2015}$/km in 2011 to 0.01 ± 0.17 EUR$_{2015}$/km in 2016. Yet, the mean differential user costs of BEVs and PHEVs remain positive, indicating that the high initial prices of these vehicles cannot, on average, yet be recovered completely during vehicle use (at the assumptions made here; see Table 1). It is important to note that user costs can vary greatly as individual cost components depend, e.g., on vehicle size and energy consumption. The average cost trend displayed by the vehicles contained in our analysis but may not, in all instances, capture accurately the user costs of each individual vehicle. The user costs of each individual vehicle are listed in Table 18 and Table 19 Appendix A.

The real user costs consist of the purchasing price, fixed yearly costs (for insurance, taxes and other fees), variable yearly costs (for maintenance, repairs, tires and replacement of other wear parts), and the costs of fuel and/or electricity. The purchasing price appears to account for the highest share in the total user costs of BEVs, PHEVs, and CVs (Figures 6 - 8). This observation might be explained by our assumption to consider a relatively short lifespan of 6 years. We find a lower share of the purchasing price in the total user costs of CVs (64% in 2016) than for BEVs (73% in 2016) and PHEVs (73% in 2016). This observation can be explained by the comparatively high price of BEVs relative to conventional vehicles. For the overall user costs as well as for the respective shares we can't find any trend for BEVs, whereas PHEVS show a slight increase in the user costs from 2011 to 2016. For CVs, the total user costs have more than doubled from 2010 to 2016 (Figure 8). This observation can be explained by the increasing purchasing price, which developed from 10,340 EUR to 40,135 EUR during this period.

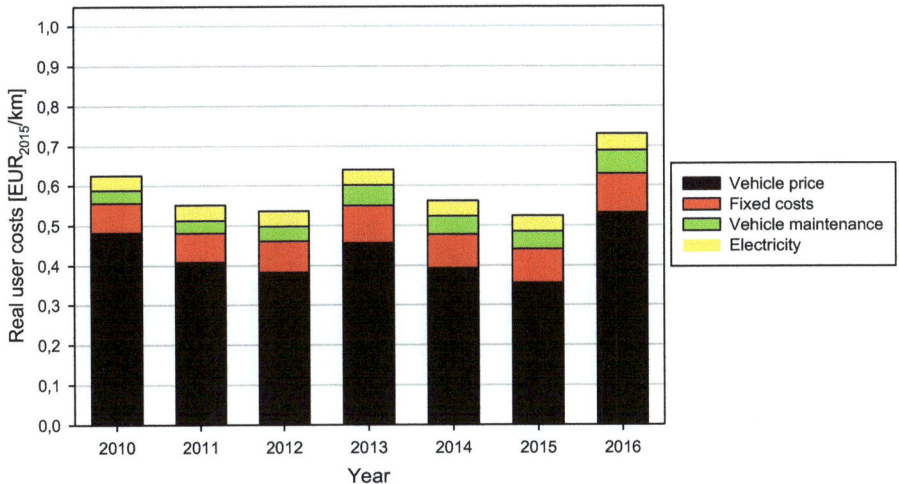

Figure 6: Real user costs of BEVs disaggregated into four principal cost components

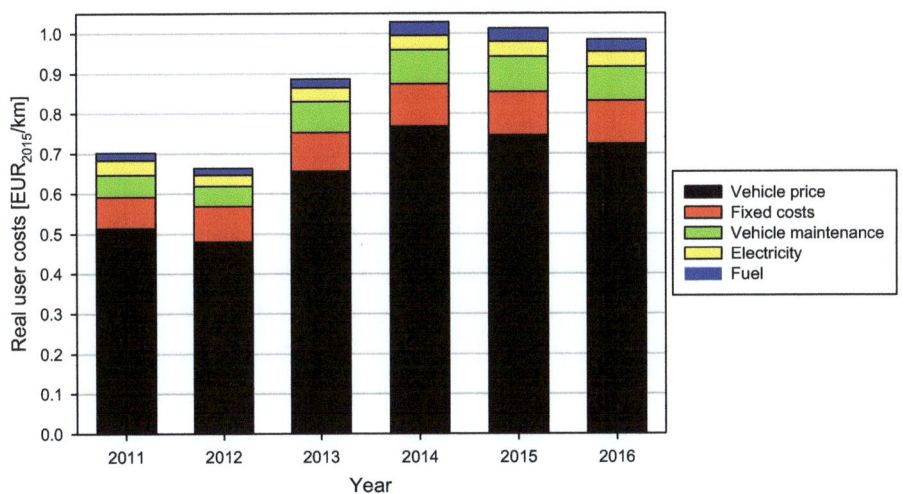

Figure 7: Real user costs of PHEVs disaggregated into five principal cost components

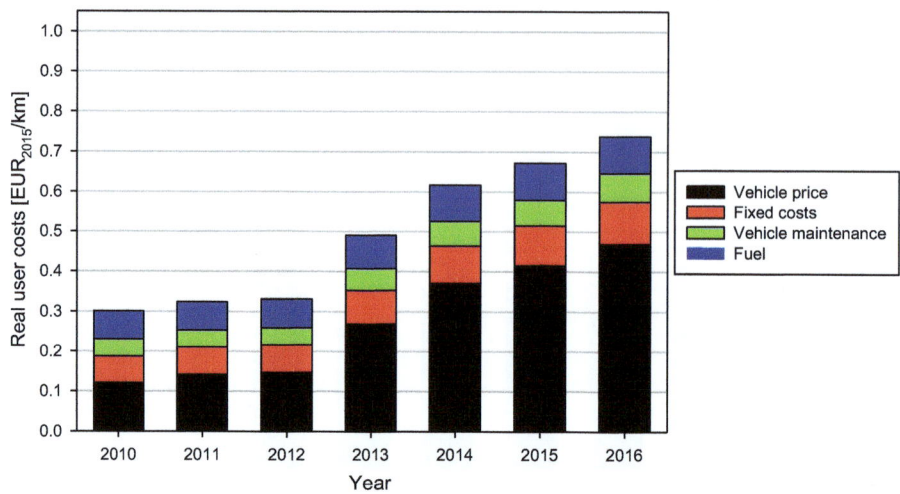

Figure 8: Real user costs of CVs disaggregated into four principal cost components

Considering the cost allocation in 2016 for BEVs, PHEVs and CVs, we find that for PHEVs the purchasing price accounts for the highest share in the real user costs (73%) whereas the fuel and electricity costs have the lowest share (7%). In contrast, for CVs, the purchasing price only accounts for 64% of the user costs. For this category, the share of the fuel costs (14%) in the total user costs is higher than for BEVs and PHEVs.

The findings are in line with other studies, which also find depreciation having the highest share of the user costs, for example 40% for normal sized conventional vehicles (VTPI, 2017). An older study of Dekra already found depreciation to be responsible for half of the user costs (DEKRA, 2008). Other studies also find higher shares of the purchasing price on the user costs for BEV and PHEV compared to conventional vehicles and also identify the purchasing price as the main cost factor (Rousseau et al., 2015). Another study finds a share of the purchasing price of 49% for vehicles newly bought in the U. S., which is lower than our findings (Hagman et al., 2015).

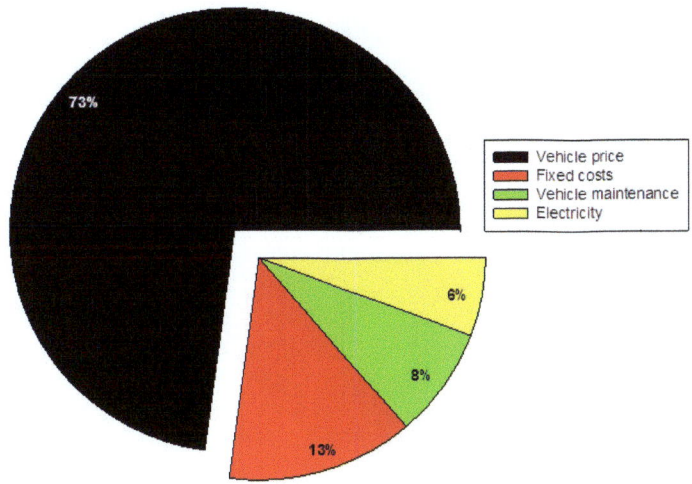

Figure 9: Share of cost components in the real user costs of BEVs in the year 2016

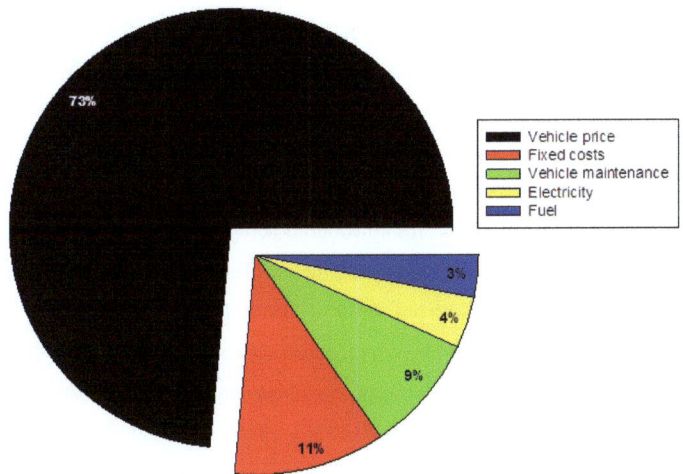

Figure 10: Share of cost components in the real user costs of PHEVs in the year 2016

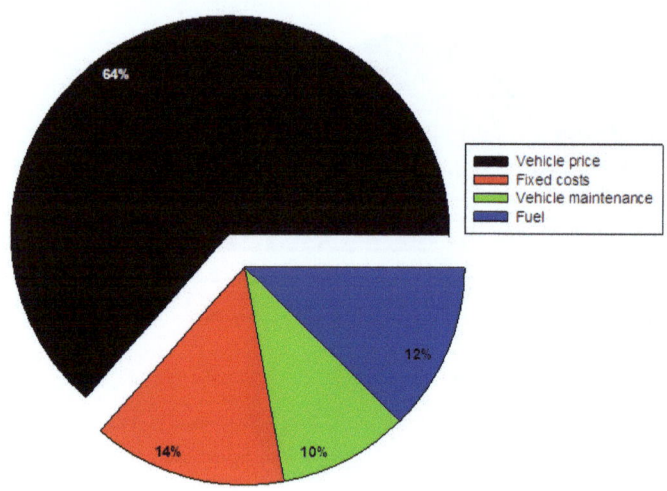

Figure 11: Share of cost components in the real user costs of CVs in the year 2016

3.4 Costs of abating carbon dioxide by BEVs and PHEVs

The CO_2 emissions of vehicles vary depending on whether one considers the certified emission levels, the actual on-road emissions, or the emissions along the entire WTW chain. The mean cost estimates are often associated with large error margins resulting from BEVs and PHEVs with comparatively high user costs but low CO_2 emission savings compared to conventional vehicles[3]. We find that the costs of emissions abatement by BEVs tend to decrease in the period from 2010 to 2016 for all four scenarios considered here (Figure 12), whereas as no uniform trend for PHEVs can be observed. Considering the certified tailpipe CO_2 emission values, we see a cost decrease for BEVs from 0.30 ± 0.12 EUR_{2015}/100 g CO_2 in 2010 to 0.15 ± 0.33 EUR_{2015}/100 g CO_2 in 2016, which constitutes a decrease of 50% (Figure 12a). For PHEVs, we observe a similar development, resulting in a cost decrease of 73% (0.14 ± 0 EUR_{2015}/100 g CO_2 in 2011 to 0.04 ± 0.10 EUR_{2015}/100 g CO_2 in 2016). The certified CO_2 emissions at the tailpipe underestimate, however, the actual amount of CO_2 [g/km] emitted when vehicles are driven on the road (e.g., ICCT, 2016b; Franco et al., 2016). Considering the actual *on-road* CO_2 emissions at the tailpipe, the mitigation costs decrease for BEVs by around 60% (from 0.24 ± 0.10 EUR_{2015}/100 g CO_2 in 2010 to 0.1 ± 0.24 EUR_{2015}/100 g CO_2 in 2016) and by 22% for PHEVs (Figure 12b). The increasing mitigation costs of the on-road CO_2 emissions of PHEVs as compared to the certified emissions could be a result of the large gap between

---

[3] Following the calculation in Equation 6, the costs of mitigating CO_2 emissions can become infinitively high, if the differences in user costs and CO_2 emissions between BEVs and PHEVs versus CVs become very large or very small, respectively.

certified and real world CO_2 emissions of PHEVs lowering the price difference between PHEVs and CVs.

The tail-pipe CO_2 emissions considered so far do not yet account for the CO_2 that is emitted during the production of electricity for BEVs and PHEVs as well as the carbon losses along the fuel-supply chain. Including these potentially important sources of CO_2 emissions inking a more comprehensive WTW perspective suggests likewise a solid decrease of the costs of CO_2 emissions reduction for BEVs (Figure 12). The average addition costs over conventional vehicles decrease by a staggering 89% from 2010 to 2016, resulting in additional costs of emission reduction of 0.12 ± 0.78 $EUR_{2015}/100$ g CO_2 in 2016. For PHEV, we see an 88% decrease in the costs of abating WTW CO_2 emissions from 2011 to 2014. For 2015 and 2016 we find negative values, meaning the emissions of PHEV are higher compared to the emissions of CVs. Compared to the on-road scenario, the WTW scenario shows a much lower difference between BEVs and PHEVs because in the former BEVs are considered with zero emissions, whereas in the latter, the CO_2 emissions of the production of the electric energy are considered.

As BEVs and *serial* PHEVs contain traction batteries of high energy storage capacity whose production is energy intensive and therefore results in CO_2 emissions, we take the emissions impact of battery production into consideration in a hybrid WTW scenario that was first proposed by Moro and Helmers (2015). For this scenario, we obtain higher costs of CO_2 emission mitigation through BEVs than for the previously considered scenarios (-13.02 ± 30.69 $EUR_{2015}/100$ g CO_2 in 2010 and 0.18 ± 2.16 $EUR_{2015}/100$ g CO_2 in 2016). For BEVs, the costs of emission reduction decreases by 91% from 2011 to 2016. For PHEVs, the costs of mitigating CO_2 emissions may even be negative, meaning that PHEVs emit more CO_2 than their conventional counterparts if battery manufacturing and the entire WTW chain is considered. Considering the entire period between 2011 and 2016, PHEVs tend to emit 79 ± 87 g CO_2/km more on average than CVs in the hybrid WTW scenario. We cannot find a trend for this difference.

The results of our four scenarios suggest that the costs of mitigating CO_2 emissions by BEVs and PHEVs increase, if additional sources of CO_2 emissions along vehicle and production are taken into consideration. For PHEV we only find decreasing costs when considering tail-pipe emissions and we cannot even find savings in the hybrid WTW scenario. The costs of mitigating CO_2 emissions in the hybrid WTW scenario translate for BEVs into $1789 \pm 21,412$ EUR_{2015}/t CO_2. This value is less than the around 2000-2500 €/t CO_2 equivalent found by ASUE

(ASUE, 2016). It compares to 345 to 570 EUR/t CO_2 for conventional heat insulation for buildings (Gambardella et al., 2012), 311 to 846 EUR/t CO_2 for photovoltaics (Blesl, 2011; Beer, 2005), 150 for off-shore wind power (Blesl, 2011), 22 natural gas-powered combined cycle power plants (Blesl, 2011), 15-30 EUR/t CO_2 for nuclear electricity generation (Fritsche, 2007). It should be considered, that CO_2 abatement costs always depend on the chosen assumptions and calculation model as well as government grants. Further, CO_2 abatement of the electricity production might also influence the abatement costs of electric vehicles, as we assume the average electricity mix in Germany for calculation.

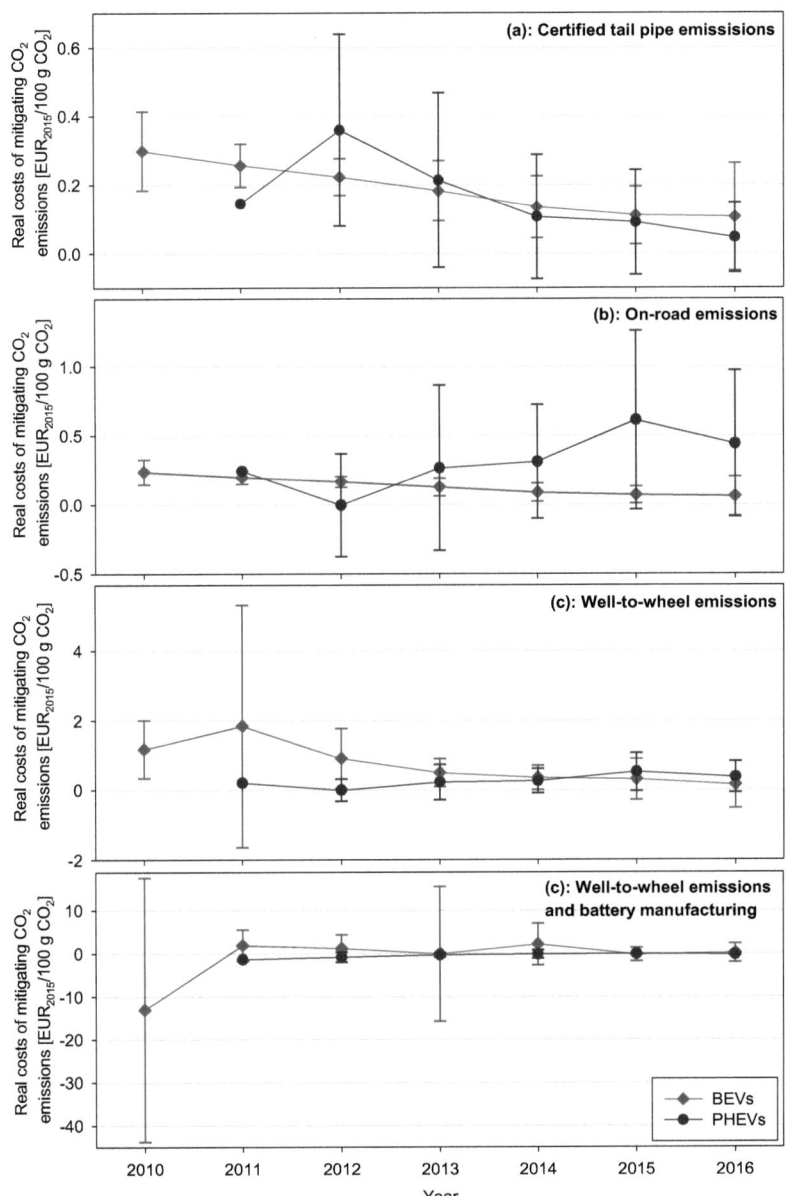

Figure 12: Real costs of mitigating the CO_2 emissions of conventional vehicles by BEVs and PHEVs considering: certified emission levels at the tailpipe (a), actual real-word tailpipe emissions on the road (b), emissions along the the entire well-to-wheel chain of electricity and fuels (c) and a hybrid approach include the emissions of the entire

well-to-wheel chain of electricity and fuels and the emissions of battery production (d); error bars represent the standard deviation of cost data

For BEV we find lowest abatement costs in the on road scenario. In comparison to the certified scenario, the on road scenario assumes higher fuel and electricity costs and therefore higher user costs. As both scenarios don't consider CO_2 emissions for BEV, the saved CO_2 increases in the on road scenario and therefore the costs per CO_2 saving decrease. For PHEVs, we assume a much higher correction factor in the on road scenario, see Table 1. As PHEVs also emit CO_2 in the certified scenario due to their combustion process, we find, in contrast to BEVs, increasing abatement costs in the on road scenario, as the savings due to the higher on road correction factor of PHEVs compared to their conventional counterparts, see Figure 13. In the WTW scenario, we find mich higher abatement costs for BEVs, as power plant CO_2 emissions are considered and therefore the CO_2 savings go down what increases the cost per saving. In the WTW scenario, we already find PHEVs to emit more CO_2 than CVs by 67 g on average, meaning there are many negative values, which falsely bring down the abatement costs for PHEVs in the WTW scenario. In the WTW hybrid scenario, we find PHEVs emitting 79 g more CO_2 on average than CVs, resulting in overall negative abatement costs. For BEVs we find in the WTW hybrid scenario lower abatement costs, however, this scenario calculation might most likely be inaccurate, because the gap in CO_2 emissions between BEVs and CVs almost completely disappears, resulting in a very high modulus of abatement costs and therefore a very high standard.

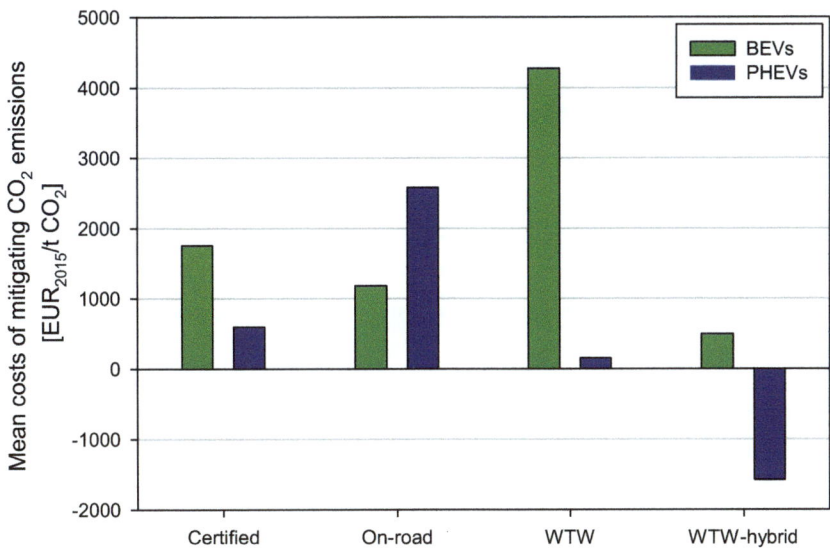

Figure 13: Mean real costs of mitigating the CO_2 emissions of conventional vehicles by BEVs and PHEVs for the four different scenarios; results are indicative only as uncertainty margins are very large (not depicted here).

3.5 Costs of mitigating NO_X and PN emissions by BEVs and PHEVs

The costs of mitigating NO_X emissions through the deployment of BEVs and PHEVs are generally lower when diesel cars rather than gasoline cars are considered as reference (Figure 144, top); this observation is related to the substantially higher on-road NO_X emissions of diesel as compared to gasoline cars. The trend towards decreasing costs of mitigating NO_X emissions from gasoline cars is similar for BEVs and PHEVs (Figure 14a). For BEVs, the real costs of NO_X emission reduction decreased by 55% from 2010 to 2016 (i.e., from 0.56 ± 0.25 EUR_{2015}/100 g NO_X in 2010 to 0.25 ± 0.60 EUR_{2015}/100 g NO_X in 2016); for PHEV, the costs decreased by 93% from 2011 to 2016 (i.e., from 0.41 ± 0 EUR_{2015}/100 g NO_X in 2011 to around 0.03 ± 0.36 EUR_{2015}/100 g NO_X in 2016). Assuming NO_X emission factors for diesel cars (Figure 14, top), the costs of BEVs decreased in a similar way by 30%. For the years 2015 and 2016 we see deviating cost values for PHEVs, caused by the lower emissions of the Euro 6 vehicles assumed as reference for these two years (see Table 2). From 2011 to 2014 we find a decrease in the mitigation costs of PHEVs of 60%.

Regarding particle number emissions, we find that the mitigation costs of BEVs decrease by 89% and 43% between 2010 and 2016 for reducing PN emissions of gasoline and diesel cars, respectively (Figure 14b, bottom). For PHEVs, the costs decrease by 97% if emission factors of gasoline cars are assumed as reference. Due to our assumptions that are based on tests of a very limited number of cars (see Table 2), diesel-propelled PHEVs could even have higher PN emissions than comparable diesel CVs. If one assumes that PHEVs may be driven electrically for at least short distances (thus not exhibiting tailpipe emissions during parts of the trip), there is *a priori* no reason to assume higher PN emissions for Euro 6 PHEVs than for conventional diesel cars[4]. We decided therefore to set all cost values to zero (Figure 14b, bottom) instead of the calculated negative values. We find a decrease of PN abatement costs for BEVs in both scenarios, most likely due the increasing user costs of CVs due to their raising purchasing price compared to the relatively constant user costs of BEVs.

In a second scenario, we consider next to tailpipe emissions also the NO_X emissions related to the electricity production. For BEVs, the costs of mitigating NO_X emissions decrease by 80% (from 3.52 ± 3.80 $EUR_{2015}/100$ g NO_X in 2010 to 0.72 ± 4.39 $EUR_{2015}/100$ g NO_X in 2016) in the gasoline scenario and by 27% (from 0.07 ± 0.03 $EUR_{2015}/100$ g NO_X in 2010 to 0.05 ± 0.09 $EUR_{2015}/100$ g NO_X in 2016) in the diesel scenario, see Figure 14. PHEVs do not mitigate NO_X emissions in comparison to their conventional counterparts from 2011 to 2015 in the gasoline scenario. For 2016 we find savings and find costs of 0.05 ± 2.93 $EUR_{2015}/100$ g NO_X for the emission reduction. For PHEVs in the diesel scenario we find decreasing costs of NO_X emission reduction from 2011 to 2014 by 66% (from 0.34 ± 0 $EUR_{2015}/100$ g NO_X to 0.12 ± 0.20 $EUR_{2015}/100$ g NO_X) but cannot find savings for 2015 and 2016, meaning PHEVs emitted more NO_X than their conventional diesel cars in these years, most likely due to the assumption of the different certified emissions standard for 2015 and 2016 (EURO 6 as compared to EURO 5 for the years between 2010 and 2014).

---

[4] Yet, the assumption of higher PN emissions for PHEVs might be conceivable as (i) the combustion engine might be operated at times at higher than average loads, thus emitting more particles, when recharging the battery and (ii) vehicle manufacturers might install smaller particulate filters in PHEVs as compared to conventional diesel cars to save costs.

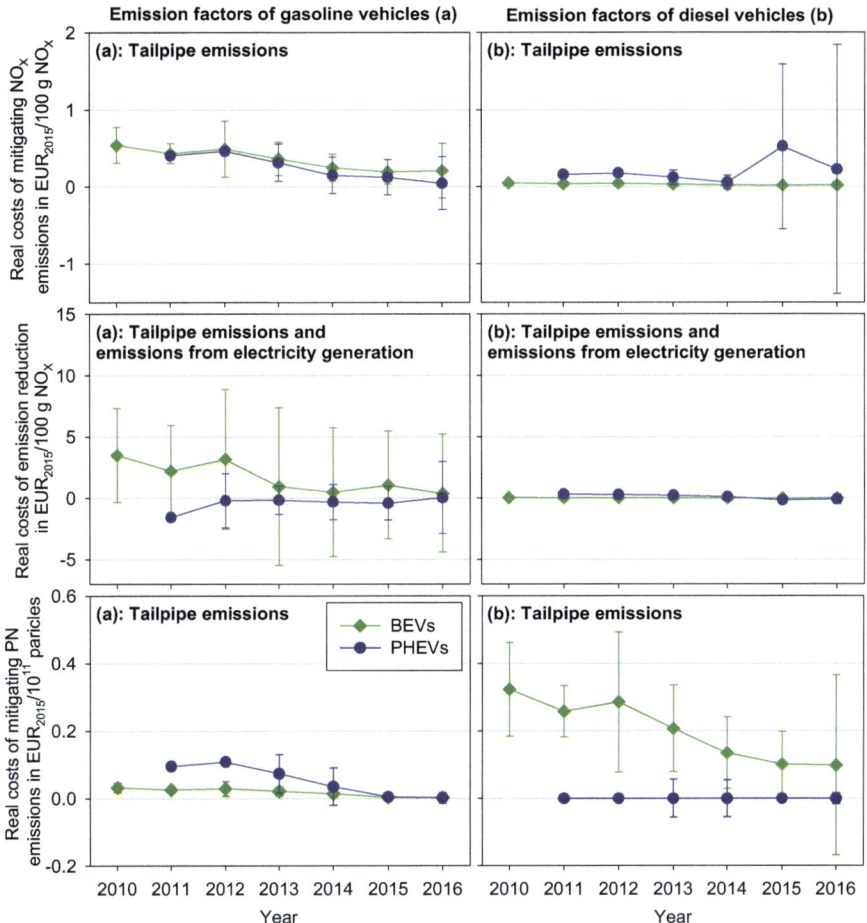

Figure 14: Real costs of mitigating the NO_X and PN emissions of conventional gasoline vehicles (a) and diesel vehicles (b) by BEVs and PHEVs considering: NO_X tailpipe emissions (top), NO_X tailpipe emissions and NO_X emissions from electricity generation (middle), and particulate number tailpipe emissions (bottom); error bars represent the standard deviation of the cost data

Regarding the costs of mitigating NO_X emissions, we obtain higher costs if all analyzed vehicles are considered as gasoline-propelled because diesel-propelled CVs have higher NO_X emissions; therefore, the savings by replacing them with BEVs or PHEVs with a gasoline engine are higher.

Considering the best-case scenarios (zero emission energy production, diesel-propelled vehicles as alternatives regarding NO_X emissions and gasoline-propelled vehicles as alternatives regarding PN emissions) BEVs mitigate NO_X emissions at costs of 530 ± 846 EUR_{2015}/t NO_X and

3729 ± 8992 EUR$_{2015}$/10^{17} particles in 2016. These costs compare to 800-3,800 EUR/t NO$_X$ for light duty vehicles and are comparable to those of mitigating NO$_X$ emissions in the manufacturing industry (see Appendix D). For reducing PN with particulate filters, mitigation costs range between 7-88 EUR/10^{17} particles for light duty vehicles.

4 Discussion

4.1 Strengths and limitations of the research

This thesis analyzes the trend in the purchasing prices for BEVs, PHEVs, and their conventional counterparts as well as the development of real user costs and the costs of mitigating CO_2 and air pollutant emissions in the period between 2010 and 2016. For the latter, we consider various scenarios that (i) capture certified tailpipe emissions, real-world on-road emissions, the entire WTW chain of fuel and electricity production, and battery manufacturing and (ii) differentiate (in the analysis of air pollutants) between emissions factors for gasoline and diesel vehicles.

We identify substantial technological learning and a robust trend towards declining prices and price differentials of BEVs and PHEVs. The user costs of both vehicle categories have been declining relative to conventional vehicles; yet, absolute user costs [EUR_{2015}/km] of BEVs, PHEVs, and CVs tend to remain constant or even increase caused by a general trend towards more powerful, thus less energy efficient vehicles and therefore inclining absolute prices. The costs of mitigating CO_2 and air pollutant emissions show considerable variability between vehicles and scenarios but tend to follow an overall declining trend for BEVs.

Our analysis quantifies for the first time technological learning and the historical trends in the costs of vehicle users as well as the costs of mitigating emissions through BEVs and PHEVs. Thereby, the thesis fills an important gap in the scientific literature that had long remained open as necessary data became available just recently. We regard our findings as robust but subject to uncertainty. The accuracy of results depends on three principal factors: (i) the reliability of the experience curve approach to quantify technological learning, (ii) the assumptions made for the calculation of user costs and the costs for mitigating CO_2 and air pollutant emissions and (iii) the scope and quality of the collected input data. The integrity of the data base on the prices and technical characteristics of vehicles is ensured by an extended web search based on different sources. Table 3 Table 6 provide a detailed overview of the most relevant strengths and limitations of the individual analysis conducted for this thesis.

Table 3: Principle strengths and limitations of this thesis

Strengths: - First experience curve analysis and cost estimates for BEVs and PHEVs based on a comprehensive set of price data
Limitations: - Geographical scope is limited to Germany   - PHEVs data are slightly inconsistent due to mix-up of PHEVs and BEVs with range extender, however, this approach is valid, because BEVs with range extender have only a minor share of around

8.5% in the analyses, lowering the average battery capacity specific price for all vehicles [€$_{2015}$/kWh] by 6.5% and the engine output specific price [EUR$_{2015}$/kW)] only by 0.1% compared to a scenario containing only the PHEVs and excluding the range extender vehicles - Prices are not equal to production costs, as the production costs account only for around two-thirds of the purchasing prices (Vyas et al., 2000); assuming purchasing prices for the experience curve analysis is valid, because production represent the major part of the prices
Scope for further research: - Learning rates and costs could be substantiated by expanding the analysis to other countries and regions; most relevant could be countries such as China, Japan, and the USA that have a large vehicle market

Table 4: Strengths and limitations of the experience curve analysis

Strengths: - Analysis demonstrates that technological learning has substantially reduced the specific prices and price differential of BEVs and PHEVs in the past - Substantial technological learning suggests further scope for cost and price reduction in the future
Limitations: - Product system is inhomogeneous as vehicles became more powerful and likely equipped with additional safety and infotainment features; both introduce biases into the learning rates identified here - Vehicle prices are used as proxy for production costs; the calculated learning rates are biased by any changes in the profit margin of vehicle manufacturers - Learning rates do only apply to the set of technologies considered here but do not allow drawing conclusions for the potentials of, e.g., alternative battery technologies - The relatively moderate price decline in recent years suggests that learning rates might not remain constant but could decrease as large parts of learning potentials might have been exhausted; this possibility might suggest a bias if our learning rates are used for mid- to long-term price forecasts of BEVs and PHEVs
Scope for further research: - Bottom-up analysis of component costs could help explaining learning rates and identify scope for technological learning in the future - Extrapolating the results of this thesis to forecast production costs and vehicle prices in the future

Table 5: Strengths and limitations of the analysis of user costs

Strengths: - The chosen lifespan of 6 years (i) is consistent with the fixed and variable maintenance costs of vehicles obtained from ADAC (2017) and (ii) matches the likely lifespan of the traction battery, which is one of the most expensive parts of electric vehicles - Maintenance, fix and energy costs are assumed as constant throughout the analysis to only image the influence of the purchasing price
Limitations: - The chosen lifespan of 6 years is equivalent to a lifetime mileage of around 86.000 km, which is substantially lower than the average lifetime mileage of vehicles in Germany (170.000 to 230.000 km over a lifespan between 14 and 18 years based on Weymar et al. (2016)); the approach made in this thesis disregards the salvage value of the vehicles after 6 years and overestimate the impact of purchasing price on the real users costs - Real user costs might fluctuate more if variable, time-dependent maintenance, fix and energy costs were assumed throughout the period under consideration - Purchasing prices are net, whereas other calculation factors of the user costs (energy costs, maintenance costs, fix costs) contain taxes. This approach is valid because the cost factors for the real user costs contain all sorts off other charges and subventions and it would be almost impossible to fully exclude all of them

- The real user costs and therefore the costs of emission reduction cannot represent the real costs like on a real, unregulated market without further influences of governance on the market prices and therefore the real user costs
- Research approach does not include a complete lifecycle scenario for the vehicles themselves but only for the production of the traction batteries
Scope for further research:
- Scenario analyses could calculate user costs for a range of plausible lifespans and yearly mileages.

Table 6: Strengths and limitation of the analysis of costs for mitigating CO2 and air pollutant emissions; note that user costs are required to calculate the costs of mitigating emissions – the strengths and limitations discussed in Table 5 also apply here

Strengths:
- First estimation of costs of mitigating CO_2 and air pollutant emissions through BEVs and PHEVs and benchmarking against alternative emission mitigation technologies
Limitations:
- Static cost calculations based on a fixed set of assumptions
- Approach does not include a complete lifecycle scenario for the vehicles themselves but only for the production of the traction batteries
Scope for further research:
- Probabilistic modeling of mitigation costs based on a range of likely user costs as well as energy use and emission scenarios
- Expansion of calculations to include the entire vehicle life cycle
- Analysis of the real costs without any taxes, fees, subventions, etc., however, it might be almost impossible to exclude all
- Further lifecycle analyses could be made in future researches

4.2 Discussion of results

The rates for technological learning of electric vehicles are influenced by the learning rates of the technical components, most importantly the traction battery, which represents the major cost component of a BEV, contributing $19 \pm 1\%$ to the overall costs of BEVs (Safari, 2017). Studies found, that the prices of lithium-ion batteries went down to one third of the previous values from 2005 to 2014 (Ciez et al., 2016); Nagelhout and Ros (2009) and Nykvist and Nilsson (2015) report learning rates for lithium-ion batteries of 17% and 6-9%, respectively. The study by Ciez et al. (2016) also suggests, that battery prices might not decrease as much as expected in the future, that could potentially slow down or even revert the robust trend towards declining BEV prices observed in this thesis. However, overall electrification costs do not only contain the costs for the battery pack, but are in fact higher. Safari finds average electrification costs of $52 \pm 2\%$ of the overall production costs of BEVs.

In the current state of the production of electric vehicles, the expense of research and development might be a significant cost factor. Decreasing battery costs might be a main reason for the decreasing real specific price of BEVs and PHEVs but there might also be non-battery-related effects of technological learning related to the integration of electric powertrain components in the vehicle and effects of mass production that may affect the cost of other vehicle components such as the chassis or infotainment systems. In fact, the learning rate of 12 ± 4% identified by Safari (2017) for the electrification cost of mid-size BEVs (excluding the traction battery) suggests substantial technological learning and thus potentials for cost reduction unrelated to the battery. Besides the battery pack, which accounts for 35-50% of the production costs, other cost components are chassis (8-20%), vehicle body (7-19%), equipment (11-27%), chassis (4-9%) as well as miscellaneous others (5-15%) (all percentage values are obtained from Kochhan et al., 2017). Assuming that the production costs account for two-thirds of the purchasing price, the battery pack as a result accounts for around 15-23% of the purchasing price. In comparison, the electric motor only accounts for 2-6%, the inverter for 2-6%. Even though the battery pack is a major cost factor, prices will decrease if other components' costs decrease.

Another research has found learning rates of 9 ± 2% for the total price and 12 ± 4% for the electrification cost of mid-size BEVs (Safari, 2017). In comparison, our research has found a learning rate of 23 ± 2 % for the engine power related real specific price [EUR_{2015}/kW] and a learning rate of 16 ± 2% for the battery capacity related real specific price. The approach made by Safari only included the "[…] closest ICV pair from the same brand […]" as a conventional counterpart and therefore excluded BEVs from the manufacturer Tesla, whereas our research includes all electric vehicles sold on the German market, even if no conventional counterpart from the same manufacturer exists. In this case, we assume similar vehicles (e. g. similar category, size, power, equipment) from other manufacturers as an equivalent. In addition, we include vehicles of all sizes and categories, as long as they are meant for normal use (e.g., no racing cars) and are produced in large quantities. As distinct from the research of Safari, we do not identify learning rates for the total price but for the engine power and battery capacity specific prices, what makes the results hardly comparable as the methods are too different. Finally, Safari (2017) does not specify to which country the price data of his analysis refer to, while our analysis is based on vehicle prices for Germany.

However, the purchasing price is not the only factor of attractiveness - the electric driving range and the charging infrastructure and in addition, the real user costs have to be considered. Publicly accessible charging points increased by around 149% from 2241 in 2011 to 5571 in 2015 in Germany but the further development is slowed down by low profitability due low utilization (NPE, 2015). The ratio of electric vehicles to charging points developed from 2 to 6.7 during the same period, meaning the amount of electric vehicles raises faster than the charging infrastructure. High costs of the traction batteries are amongst other factors responsible for the price difference between electric and conventional vehicles. Like mentioned in Section 3, the battery prices might not drop as much as expected in future due to recent studies. However, the prices of electric vehicles drawn near to the prices of the conventional vehicles, making them possibly competitive even without drastic price drops of batteries, making it more important to focus on the improvement of other key factors like charging infrastructure and driving range.

This thesis does not analyze in detail the drivers behind the price reductions of BEVs and PHEVs as we do not split the prices into their components but instead use the purchasing price as approximation of the total manufacturing costs. The approaches used in this thesis could be used to identify learning rates of BEVs and PHEVs sold on other countries' markets. The results could be compared to this research to identify possible differences in the price strategies of the manufacturers for different markets.

5 Conclusions and recommendations

Based on our findings, we draw the following conclusions:

- Technological learning has substantially reduced the specific price and price differentials of BEVs and PHEVs since 2010. As the purchasing price accounts for three fourth of the real user costs, these vehicles have become both cheaper and more cost competitive relative to conventional vehicles.

- BEVs and PHEVs mitigate CO_2 emissions at declining costs that are, however, higher than those of other technologies, e.g., insulation for buildings, wind power, or photovoltaic. The mitigation costs of BEVs and PHEVs can be decreased if batteries are produced and vehicles are powered by low-carbon renewable energy sources. Yet, we observe that part of the cost and CO_2 emission savings are already to date absorbed by a trend towards larger, more powerful vehicles, suggesting that policy makers should be aware of rebound effects that decrease the effectiveness of BEVs and PHEVs in addressing sustainability shortfalls of passenger road transport.

- BEVs have the ability to reduce CO_2 and pollution emissions, even in WTW scenarios and even when considering the emissions impact of battery production, making it worth to promote and support the electrification of the individual passenger transport. It should be considered, that prices already decreased substantially; policy makers might direct their focus to other key factors for the acceptance and attractiveness of electric vehicles, most importantly the expansions of a recharging infrastructure and electric drive range.

- The substantial technological learning observed for the specific vehicle price per unit of battery capacity is an enabler for increasing engine power (which, in turn, increases electricity consumption) and drive range of BEVs at no additional costs compared to the past cost situation. Both may make BEVs more attractive for consumers; specifically the latter allows mitigating an important shortcoming hampering the market penetration of BEVs.

- When comparing electric and conventional vehicles regarding their pollutant emissions, it should be considered, that the latter often emit pollutants harmful to health in highly crowded urban areas, whereas electricity production related emissions are mostly emitted in uninhabited areas, what represents, even under the hypothetical assumption of equivalent pollutant emissions, a meaningful advantage of electric vehicles over vehicles with internal combustion engine.

References

ADAC (2017), ADAC Autokosten Frühjahr/Sommer 2017, 17.03.1000 – IN 29021 – Stand 04-2017, https://www.adac.de/_mmm/pdf/autokostenuebersicht_47085.pdf. Retrieved: 20 January 2017

ASUE Arbeitsgemeinschaft für sparsamen und umweltfreundlichen Energieverbrauch e.V. (2016), CO_2-Vermeidung, http://www.asue.de/sites/default/files/asue/themen/umwelt_klimaschutz/2016/broschueren/07_01_16_asue_co2-vermeidung_01.pdf. Retrieved: 26 May 2017

BDEW, Bundesverband der Energie- und Wasserwirtschaft e.V. (2017), BDEW-Strompreisanalyse Februar 2017, https://www.bdew.de/internet.nsf/res/9729C83961C37094C12580C9003438D3/$file/170216_BDEW_Strompreisanalyse_Februar2017.pdf. Retrieved: 20 February 2017

Beer (2005), CO_2-Vermeidungskosten erneuerbarer Energietechnologien https://www.ffe.de/download/kurzberichte/KF_vermk.pdf. Retrieved 26 May 2017

Bischof, C. Boger, T., Gunasekaran, N. Bhargava, R. (2012): Advanced particulate filter technologies for direct injection gasoline engine applications. Corning. DEEF Conference, 16-19 October 2012. Source: https://energy.gov/sites/prod/files/2014/03/f8/deer12_bischof.pdf. Retrieved: 26 May 2017

Blesl (2011), Uni Stuttgart IER, Stand 2011, in Sonderdruck aus VBI 1/2 2013, Trippe, J., Energieeffizienz in Mittelstand und Industrieunternehmen, http://www.tpi-online.de/tl_files/pdfs/SD_Trippe.pdf. Retrieved 26 May 2017

BMF, Bundesministerium der Finanzen (2017), Afa-Tabellen, http://www.bundesfinanzministerium.de/Content/DE/Standardartikel/Themen/Steuern/Weitere_Steuerthemen/Betriebspruefung/AfA-Tabellen/afa-tabellen.html. Retrieved: 20 January 2017

BMWi (2017): Rahmenbedingungen und Anreize für Elektrofahrzeuge und Ladeinfrastruktur. BMWi – Bundesministerium für Wirtschaft und Energie. Berlin, Germany. Source: http://www.bmwi.de/DE/Themen/Industrie/Elektromobiltaet/rahmenbedingungen-und-anreize-fuer-elektrofahrzeuge.html. Retrieved: 20 January 2017

BR (2016: Leitmarkt und Leitanbieter für Elektromobilität. BR-Bundesregierung. Berlin, Germany. Source: https://www.bundesregierung.de/Webs/Breg/DE/Themen/ Energiewende/Mobilitaet/podcast/_node.html. Retrieved: 19 January 2017

Cames M, Helmers E. Critical evaluation of the European diesel car boom – global comparison, environmental effects and various national strategies. Environ Sci Eur. 2013;25(15):22. http://www. enveurope.com/content/pdf/2190-4715-25-15.pdf. Retrieved: 21 February 2017

Cames, M., Helmers, E. (2013): Critical evaluation of the European diesel car boom - global comparison, environmental effects and various national strategies. Environmental Sciences Europe 25(15), pp. 1-22.

Carslaw, D. C., Beevers, S. D., Tate, J. E., Westmoreland, E. J., Williams, M. L. (2011): Recent evidence concerning higher NO_X emissions from passenger cars and light duty vehicles. Atmospheric Environment 45, 7053-7063.

Chen, Y., Borken-Kleefeld, J. (2014): Real-driving emissions from cars and light commercial vehicles – Results from 13 years remote sensing at Zurich/CH. Atmospheric Environment 88, pp. 157-164.

Chen, Y., Borken-Kleefeld, J. (2016): NO_X emissions from diesel passenger cars worsen with age. Environmental Science and Technology 50, pp. 3327-3332.

Ciez et al. (2016), Comparison between cylindrical and prismatic lithium-ion cell costs using a process based cost model, Journal of Power Sources 340 (2017) 273e281

Degraeuwe, B., Thunis, P., Clappier, A., Weiss, M., Lefebvre, W., Janssen, S., Vranckx, S. (2016): Impact of passenger car NO_X emissions and NO_2 fractions on urban NO_2 pollution – Scenario analysis for the city of Antwerp, Belgium. Atmospheric Environment 126, pp. 218-224.

Delphi (2015): Worldwide emission standards – Passenger cars and light duty vehicles. Delphi. Source: http://delphi.com/docs/default-source/catalogs/delphi-worldwide-emissions-standards-pc-ldv-15-16.pdf. Retrieved: 26 May 2017

Delphi (2017): Worldwide emission standards – Heavy duty and off-highway vehicles. Delphi. Source: http://delphi.com/docs/default-source/worldwide-emissions-standards/2016-2017-heavy-duty-amp-off-highway-vehicles.pdf?status=Temp&sfvrsn=0.03636262961639791. Retrieved: 29 May 2017.

DHZ (2016): Rußpartikelfilter nachrüsten: Fördergeld beantragen. DHZ – Deutsche Handwerks Zeitung. Source: http://www.deutsche-handwerks-zeitung.de/russpartikelfilter-foerderung-kommt-ab-2015/150/3097/256319. Retrieved: 26 May 2017.

Die Bundesregierung, (2017), Energiewende, Neue Kraftstoffe und Antriebe – sauber und kostengünstig, https://www.bundesregierung.de/Webs/Breg/DE/Themen/Energiewende/Mobilitaet/mobilitaet_zukunft/_node.html. Retrieved: 13 June 2017.

Diez, W. (2008): Das Management der Cost-of-Ownership, eine Studie im Auftrag der DEKRA, Stuttgart 2008

Dulac, J. (2012): Global transport outlook to 2050 – Targets and scenarios for a low-carbon transport sector. International Energy Agency. Paris, France. Source: https://www.iea.org/media/workshops/2013/egrdmobility/DULAC_23052013.pdf Retrieved: 19 January 2017

E. Helmers, J. Dietz, S. Hartard (2017) Electric Car Life Cycle Assessment Based on Real-world Mileage and the Electric Conversion Scenario International Journal of Life Cycle Assessment. Published online 7-2015, printed Vol. 22, (1), pp 15–30 in conjunction with Umweltbundesamt. [Balance of emissions avoided by renewable energy sources in the year 2013] Emissionsbilanz erneuerbarer Energieträger. Bestimmung der vermiedenen Emissionen im Jahr 2013. Report 29/2014. 2014. sites/default/files/medien/378/publikationen/climate_change_29_2014_schrempf_komplett_10.11.2014_0.pdf. Retrieved: 20 February 2017

EC (2007): Regulation 715/2007. Official Journal of the European Union L171, pp. 1-16. EC – European Commission. Brussels, Belgium.

EC (2011): Roadmap to a single European transport area – Towards a competitive and resource efficient transport system. White paper COM (2011) 144 final. EC – European Commission. Brussels, Belgium.

EEA (2016): Air quality in Europe – 2016 report. EEA Report No. 28/2016. EEA- European Environment Agency. Copenhagen, Denmark.

EEA (2016): EMEP/EEA air pollutant emission inventory guidebook 2016. Section 1.A.3.b.i-iv road transport - update Dec. 2016. EEA – European Environmental Agency. Source:

http://www.eea.europa.eu/publications/emep-eea-guidebook-2016. Retrieved: 13 February 2017

EPA (2015): Assessment of non-EGU NO_X emission controls, cost of controls, and time for compliance. EPA – U.S. Environmental Protection Agency. Source: https://www.epa.gov/sites/production/files/2015-11/documents/assessment_of_non-egu_nox_emission_controls_and_appendices_a_b.pdf. Retrieved: 29 may 2017.

Eurostat (2017), European Commission Database, Prices (prc), HICP (2015 = 100) – monthly data (annual rate of change) (prc_hicp_manr), http://ec.europa.eu/eurostat/data/database?node_code=prc_hicp_manr. Retrieved: 29 May 2017

FG (2017): Rußpartikelfilter: Funktion und Haltbarkeit. FG – FairGarage. Source: https://www.fairgarage.de/russpartikelfilter. Retrieved: 25 May 2017

Franco, V., Kousoulidou, M., Muntean, M. Ntziachristos, L., Hausberger, S., Dilara, P. (2013): Road vehicle emission factors development: A review. Atmospheric Environment 70, pp. 84-97.

Franco, V., Zacharopoulou, T., Hammer, J., Schmidt, H., Mock, P., Weiss, M., Samaras, Z. (2016): Evaluation of exhaust emissions from three diesel-hybrid cars and simulation of after-treatment systems for ultralow real-world NO_X emissions. Environmental Science and Technology 50, pp. 13151-13159.

Fritsche (2007), Treibhausgasemissionen und Vermeidungskosten der nuklearen, fossilen und erneuerbaren Strombereitstellung, Öko-Institut e.V., http://www.solarenergie-zuerisee.ch/o.k/Studien%20und%20Berichte/Treibhausemissionen%20und%20Vermeidungskosten.pdf. Retrieved 26 May 2017

Gambardella et al. (2012), Vergleich der C02-Vermeidungskosten zwischen konventionellem Wärmeschutz und einem Hausautomationssystems, Ergebnisse aus dem Forschungsprojekt Connected Energy – SHAPE (AP6, Bericht D 6.2), https://www.borderstep.de/wp-content/uploads/2014/07/Gambardella-Bergset-Beucker-SHAPE_D_6.2._CO2_Vermeidungskosten-2012.pdf. Retrieved 26 May 2017

Giechaskiel, B. (2017): personal communication.

Giechaskiel, B., Riccobono, F., Mendoza, P., Grigoratos, T. (2016): Particle number (PN) – Portable emissions measurement systems (PEMS). Heavy duty vehicles evaluation phase that the Joint Research Centre (JRC). Report EUR 28256 EN. European Commission, Joint Research Centre. Ispra, Italy.

Giechaskiel, B., Riccobono, F., Vlachos, T., Mendoza-Villafuerte, P., Suarez-Bertoa, R., Fontaras, G., Bonnel, P., Weiss, M. (2015): Vehicle emission factors of solid nanoparticles in the laboratory and on the road using Portable Emission Measurement Systems (PEMS). Frontiers in Environmental Science 3:82. doi: 10.3389/fenvs.2015.00082.

Hagman et al. (2016), Total cost of ownership and its potential implications for battery electric vehicle diffusion, Research in Transportation Business & Management Volume 18, March 2016, Pages 11–17 http://www.sciencedirect.com/science/article/pii/S2210539516000043. Retrieved: 20 May 2017

Hammer, J., Schmidt, H., Franco, V., Posada Sánchez, F., Samaras, Z., Zacharopoulou, T. (2015): Development of a method for assessing real-world emissions of hybrid diesel light duty vehicles. Draft Final Report. TÜV Nord, ICCT, LAT.

Helmers, E. (2010): Bewertung der Umwelteffizienz moderner Autoantriebe – Auf dem Weg vom Diesel-Pkw-Boom zu Elektroautos. Umweltwissenschaften und Schadstoff-Forschung 22(5), pp. 564-578.

Helmers, E., Weiss, M. (2017): Advances and critical aspects in the life-cycle assessment of battery electric cars. Energy and Emission Control Technologies 5, pp. 1-18.

Hoffman (2014), orig. source PV Magazine, Stored electricity heads to $0.05/kWh by 2030 http://reneweconomy.com.au/stored-electricity-heads-to-0-05kwh-by-2030-39889/. Retrieved: 20 May 2017

Huang, C., Chen, C. H., Li, L., Cheng, Z., Wang, H. L., Huand, H. Y., Streets, D. G., Wang, Y. J., Zhang, G. F., Chen, Y. R. (2011): Emission inventory of anthropogenic air pollutants and VOC species in the Yangtze River Delta region, China. Atmospheric Chemistry and Physics 11, pp. 4105-4120.

IA-HEV (2015): Hybrid and electric vehicles. IA-HEV – Implementing Agreement for Co-operation on Hybrid and Electric Vehicle Technologies and Programmes. Cited from IEA (2016b).

ICCT (2016): NO_X emissions from heavy-duty and light-duty diesel vehicles in the EU: Comparison of real-world performance and current type-approval requirements. ICCT – The

International Council on Clean Transportation. Source: http://www.theicct.org/sites/default/files/publications/Euro-VI-versus-6_ICCT_briefing_06012017.pdf. Retrieved: 29 May 2017

ICCT (2016b), Real-world fuel consumption and CO_2 Emissions of new passenger cars in Europe, http://www.theicct.org/laboratory-road-2016-update. 20 January 2017

IEA (2016a): International Energy Outlook. IEA – United States Energy Information Administration. Washington, USA. Source: https://www.eia.gov/outlooks/ieo/pdf/transportation.pdf. Retrieved: 4 January 2017

IEA (2016b): Global EV outlook 2016. IEA – International Energy Agency. Paris, France. Source: https://www.iea.org/publications/freepublications/publication/Global_EV_Outlook_2016.pdf. Retrieved: 20 January 2017

Kahn R, Kobayashi SS, Beuthe M., et al. (2007) Transport and its infrastructure. In: Metz B, Davidson OR, Bosch PR, Dave R, Meyer LA, eds. Climate change 2007: mitigation contribution of working group III to the fourth assessment report of the intergovernmental panel on climate change. Cambridge and New York: Cambridge University Press

KBA, Kraftfahr-Bundesamt (2015), Pressemitteilung Nr. 15/2015, https://www.kba.de/SharedDocs/Pressemitteilungen/DE/2015/pm_15_15_jaehrliche_fahrleistung_deutscher_pkw_pdf.pdf?__blob=publicationFile&v=5. Retrieved: 20 April 2017.

Kochhan et al. (2014), An Overview of Costs for Vehicle Components, Fuels and Greenhouse Gas Emissions, https://www.researchgate.net/publication/260339436_An_Overview_of_Costs_for_Vehicle_Components_Fuels_and_Greenhouse_Gas_Emissions. Retrieved 27 May 2017

Masiero, G., Ogasavara, M. H., Jussani, A. C., Risso, M. L. (2016): Electric vehicles in China: BYD strategies and government subsidies. RAI – Revista de Administração e Inovação 13, pp. 3-11.

Maycock (2014), BLOOMBERG NEW ENERGY FINANCE SUMMIT 2015, Bloomberg New Energy Finance, Battery University, MIIT, https://data.bloomberglp.com/bnef/sites/4/2015/04/BNEF-2015-Keynote-ML.pdf. Retrieved: 20 February 2017

Minjares, R. J., Posada Sanchez, F. (2011): Estimated cost of gasoline particulate filters. Working paper 2011-8. ICCT – The International Council on Clean Transportation. Source: http://www.theicct.org/sites/default/files/publications/GFPworkingpaper2011.pdf. Retrieved: 26 May 2017

Mooney, J. J. (2007): The 3-was catalytic converter. California Air Resources Board Chairman Invitational Seminar Series. Source: https://www.arb.ca.gov/research/seminars/mooney/mooney.pdf. Retrieved: 26 May 2017.

Moro, A., Helmers, E. (2015): A new hybrid method for reducing the gap between WTW and LCA in the carbon footprint assessment of electric vehicles. International Journal of Life Cycle Assessment 22(1), pp. 4-14.

Nagelhout, D., Ros, J. P. M. (2009): Elektrisch autorijden – Evaluatie van transities op basis van systemopties. Report 500083010. PBL – Planbureau voor de Leefomgeving. Bilthoven, The Netherlands.

NPE, Verfasser Nationale Plattform Elektromobilität (2015), Ladeinfrastruktur für Elektrofahrzeuge in Deutschland, Statusbericht und Handlungsempfehlungen 2015, Gemeinsame Geschäftsstelle Elektromobilität der Bundesregierung (GGEMO), http://nationale-plattform-elektromobilitaet.de/fileadmin/user_upload/Redaktion/NPE_AG3_Statusbericht_LIS_2015_barr_bf.pdf. Retrieved 27 May 2017

Nykvist, B., Nilsson, M. (2015): Rapidly falling costs of battery packs for electric vehicles. Nature Climate Change 5, pp. 329-332.

OECD/IEA 2016, Global EV Outlookhttps://www.iea.org/publications/freepublications/publication/Global_EV_Outlook_2016.pdfECB 2017 https://www.ecb.europa.eu/stats/prices/hicp/html/index.en.html 1-4-2017

Posada, F., Bandivadekar, A., German, J. (2012): Estimated cost of emission reduction technologies for light-duty vehicles. ICCT – The International Council on Clean Transportation. Source: http://www.theicct.org/sites/default/files/publications/ICCT_LDVcostsreport_2012.pdf. Retrieved: 26 May 2017

Posada, F., Chambliss, S., Blumberg, K. (2016): Costs of emission reduction technologies for heavy-duty diesel vehicles. ICCT White Paper. ICCT- The International Council on

Clean Transportation. Source: http://www.theicct.org/sites/default/files/publications/ICCT_costs-emission-reduction-tech-HDV_20160229.pdf. Retrieved: 29 May 2017

Rousseau et al. (2015), Comparison of Energy Consumption and Costs of Different Plug-in Electric Vehicles in European and American Context, oai:elib.dlr.de:96846, Institute of Transport Research:Publications, http://elib.dlr.de/96846/1/EVS28_Task25_submitted.pdf. Retrieved: 25 May 2017

Safari, M. (2017): Battery electric vehicles: Looking behind to move forward. Submitted for publication in Energy Policy.

SB, Statistisches Bundesamt; MWV; Energie Informationsdienst, https://de.statista.com/statistik/daten/studie/779/umfrage/durchschnittspreis-fuer-dieselkraftstoff-seit-dem-jahr-1950/ https://de.statista.com/statistik/daten/studie/776/umfrage/durchschnittspreis-fuer-super-benzin-seit-dem-jahr-1972/. Retrieved: 21 February 2017.

SC (2012) Energy saving and new energy auto industry development plan (2012-2020). SC – State Council. Source: http://www.gov.cn/zwgk/2012-07/09/content_2179032.htm Retrieved: 20 January 2017. Cited from: de Neve, P. A. (2014): Electric vehicles in China. Belfer Center Policy Brief. Harvard University. Cambridge, USA.

Statista (2017), Entwicklung des Mehrwertsteuersatzes in Deutschland von 1968 bis 2017, https://de.statista.com/statistik/daten/studie/164066/umfrage/entwicklung-des-mehrwertsteuersatzes-in-deutschland-ab-1968/. Retrieved: 4. January 2017

Tietge et al. (2015), icct, From laboratory to roa - A 2015 update of official and "real-world" fuel consumption and CO_2 values for passenger cars in Europe http://www.theicct.org/sites/default/files/publications/ICCT_LaboratoryToRoad_2015_Report_English.pdf. Retrieved: 29 May 2017

TWH (2016): Fact sheet: Obama administration announces federal and private sector actions to accelerate electric vehicle adoption. TWH – The White House. Washington, USA. Source: https://www.whitehouse.gov/the-press-office/2016/07/21/fact-sheet-obama-administration-announces-federal-and-private-sector. Retrieved: 19 January 2017

UNECE (2016): Proposal for amendments to global technical regulation No. 15 on Worldwide harmonized Light vehicles Test Procedure (WLTP). UNECE – United Nations Economic Commission for Europe. ECE/TRANS/WP.29/GRPE/2016/3. Seventy-second session 12-15 January 2016. Geneva, Switzerland.

VTPI, Victoria Transport Policy Institute (2017), Transportation Cost and Benefit Analysis II – Vehicle Costs, http://www.vtpi.org/tca/tca0501.pdf. Retrieved: 20 May 2017

Vyas et al. (2000), Comparison of indirect cost multipliers for vehicle manufacturing, http://www.ipd.anl.gov/anlpubs/2000/05/36074.pdf. Retrieved 27 May 2017

Weiss, M., Bonnel, P., Kühlwein, J., Provenza, A., Lambrecht, U., Alessandrini, S., Carriero, M.,Colombo, R., Forni, F., Lanappe, G., Le Lijour, P. (2012): Will Euro 6 reduce the NO_X emissions of new diesel cars? – Insights from on-road tests with Portable Emissions Measurement Systems (PEMS). Atmospheric Environment 62, pp. 657-665.

Weiss, M., Dekker, P., Moro, A., Scholz, H., Patel, M.K. (2015): On the electrification of road transportation – A review of the environmental, economic, and social performance of electric two-wheelers. Transportation Research Part D: Transport and Environment 41, pp. 348-366.

Weiss, M., Junginger, M., Patel, M. K., Blok, K. (2010): A review of experience curve analyses for energy demand technologies. Technological Forecasting and Social Change 77, pp. 411-428.

Weiss, M., Patel, M.K., Junginger, M., Perujo, A., Bonnel, P. (2012): Learning rates and price projections for hybrid-electric and battery-electric vehicles. Energy Policy 48, pp. 374-393.

Weymar et al. (2016)., Statistical analysis of empirical lifetime mileage data for automotive LCA, Int J Life Cycle Assess (2016) 21:215–223 DOI 10.1007/s11367-015-1020-6, http://www.ipd.anl.gov/anlpubs/2000/05/36074.pdf. Retrieved 27 May 2017

WHO (2016): WHO releases country estimates on air pollution exposure and health impact. WHO – World Health Organization, Geneva, Switzerland. Source: http://www.who.int/mediacentre/news/releases/2016/air-pollution-estimates/en/. Retrieved: 20 January 2017

Yang, L., Franco, V., Mock, P., Kolke, R., Zhang, S., Wu, Y., German, J. (2015): Experimental assessment of NO$_X$ emissions from 73 Euro 6 diesel passenger cars. Environmental Science and Technology 49 (24), pp. 14409-14415.

Zerfass (2015), Energieverbrauch von Elektroautos unter Realbedingungen (Bachelor-Thesis) Umwelt-Campus Birkenfeld, https://www.dropbox.com/s/q3b60jft3pf1eyw/Bachelor-Thesis-A-Zerfass%20%282015%29.pdf?dl=0

ZSW (2016), Zentrum für Sonnenenergie- und Wasserstoff-Forschung Baden-Württemberg, Zahl der Elektroautos weltweit auf 1,3 Millionen gestiegen, https://www.zsw-bw.de/fileadmin/user_upload/PDFs/Pressemitteilungen/2016/pi05-2016-ZSW-ZahlenElektromobilitaet_01.pdf. Retrieved: 29 May 2017

Sources for vehicle-data are various car manufacturers' websites as well as and newspaper articles and other websites like Autobild, Auto-Motor-Sport, Autozeitung, goingelectric, Focus, Spiegel amongst many others, all websites retrieved until 31 January 2017, see following lists:

List 1: BEVs and equivalent CVs – purchasing price data

Mitsubishi   i-MiEV   2010   Daten: Mitsubishi i-MiEV. (Nicht mehr online verfügbar.) MITSUBISHI MOTORS Deutschland GmbH, ehemals im Original, abgerufen am 10. Januar 2011: „sinnvolle Innovationen für eine saubere Welt"   Citroën C1   http://www.sueddeutsche.de/auto/praxistest-citron-c-der-kleine-gallier-1.829491

Fiat   Karabag 500 E   2010   https://www.adac.de/_ext/itr/tests/Autotest/AT4452_Karabag_500_E/Karabag_500_E.pdf   Fiat 500   http://www.auto-motor-und-sport.de/dauertest/fiat-500-1-2-pop-der-kleinwagen-im-50-000-km-dauertest-1792854.html

German E-Cars   Stromos   2010   http://www.elektroauto-gebraucht-kaufen.de/schnaeppchen/german-e-cars-gebraucht-schnaeppchen/   Opel Agila B / Suzuki Splash 2010   http://www.opel-infos.de/infomaterial/2_13_2010-04-26_1.pdf

Tazzari Zero   2010   http://www.autobild.de/artikel/fahrbericht-tazzari-zero-1124108.html   http://www.auto-motor-und-sport.de/fahrberichte/e-auto-tazzari-zero-im-test-e-auto-mit-hohem-gefahrenpotenzial-2785833.html   Citroën C1   http://www.sueddeutsche.de/auto/praxistest-citron-c-der-kleine-gallier-1.829491

Mitsubishi   i-MiEV   2011   Daten: Mitsubishi i-MiEV. (Nicht mehr online verfügbar.) MITSUBISHI MOTORS Deutschland GmbH, ehemals im Original, abgerufen am 10. Januar 2011: „sinnvolle Innovationen für eine saubere Welt"   Citroën C1   http://www.sueddeutsche.de/auto/praxistest-citron-c-der-kleine-gallier-1.829491

Renault Kangoo Z.E   2011   http://www.elektroauto-news.net/elektroautos/renault/renault-kangoo-ze-elektro-transporter-fuer-die-city excepted (no price with battery)

Fiat   Karabag 500 E   2011   http://www.green-motors.de/auto/karabag-500-e   Fiat 500   http://www.eurocars4you.de/index.php?option=com_content&view=article&id=73&Itemid=92

Aixam Mega e-City 2011 http://www.motor-talk.de/news/elektro-auto-fahrbericht-mega-e-city-t3348778.html Aixam / Ligier Annahme basierend auf sonstigen Preisen und dieser Aussage http://derstandard.at/1271374993568/SCHWERPUNKT-STADTAUTOS-Mopedauto-als-Alternative-fuer-die-Stadt

German E-Cars Stromos 2011 http://www.elektroauto-gebraucht-kaufen.de/schnaeppchen/german-e-cars-gebraucht-schnaeppchen/ Opel Agila B / Suzuki Splash http://www.opel-infos.de/infomaterial/2_13_2011-11-28_1.pdf

Tazzari Zero 2011 http://www.autobild.de/artikel/fahrbericht-tazzari-zero-1124108.html http://www.auto-motor-und-sport.de/fahrberichte/e-auto-tazzari-zero-im-test-e-auto-mit-hohem-gefahrenpotenzial-2785833.html Citroën C1 http://www.sueddeutsche.de/auto/praxistest-citron-c-der-kleine-gallier-1.829491

German E-Cars CETOS 2012 http://www.green-motors.de/auto/german-e-cars-cetos Opel Corsa http://www.opel-infos.de/infomaterial/20_13_2012-06-18_1.pdf

Citroën C-Zero 2012 http://www.focus.de/auto/elektroauto/citroen-c-zero-kleine-rappelkiste-citroens-mini-stromer_id_3729694.html Citroën C1 http://www.sueddeutsche.de/auto/fahrbericht-citroen-c-style-der-preis-ist-heiss-1.552494

Smart ED 2012 http://www.goingelectric.de/2012/06/11/news/smart-fortwo-electric-drive-preis/ Smart fortwo http://blog.mercedes-benz-passion.com/2012/02/smart-fortwo-facelift-2012-alle-details-und-preise/

PG Elektrus 2012 http://www.mein-elektroauto.com/2012/01/deutsches-elektrauto-pg-elektrus-ist-ein-echter-sportwagen/4453/ Mercedes-Benz SLS AMG Roadster http://www.n-tv.de/auto/Praxistest-Mercedes-SLS-AMG-Roadster-article7179891.html

Renault Fluence Z.E. 2012 http://www.zeit.de/auto/2011-11/renault-fluence excepted (no price with battery)

Peugeot iOn 2012 http://www.autobild.de/artikel/peugeot-ion-preis-2773941.html Citroën C1 http://www.sueddeutsche.de/auto/fahrbericht-citroen-c-style-der-preis-ist-heiss-1.552494

Mitsubishi i-MiEV 2012 Mitsubishi Österreich: Produktseite des i-MiEV (Memento vom 4. Januar 2014 im Internet Archive) // MITSUBISHI MOTORS Deutschland Website des i-MiEV Citroën C1 http://www.sueddeutsche.de/auto/fahrbericht-citroen-c-style-der-preis-ist-heiss-1.552494

Renault Kangoo Z.E 2012 http://www.green-motors.de/auto/renault-kangoo-ze excepted (no price with battery)

Fiat Karabag 500 E 2012 http://www.green-motors.de/auto/karabag-500-e Fiat 500 http://www.autohaus-raber.de/tl_files/pdf/Fiat_500_150_PL.pdf

Nissan Leaf Visia 2012 http://www.focus.de/auto/fahrberichte/praxistest-nissan-leaf-technische-daten-nissan-leaf-2014_id_3695679.html http://www.goingelectric.de/elektroautos/nissan-leaf-2012/ Ford Focus 1.0 EcoBoost file://2012_09_05.pdf

German E-Cars Stromos 2012 http://www.elektroautomobile.org/app/download/6343536056/Kurzpreisliste+PKW+2012.pdf?t=1355130875 Opel Agila B / Suzuki Splash http://www.opel-infos.de/infomaterial/2_13_2012-11-19_1.pdf

Tazzari Zero 2012 http://www.autobild.de/artikel/fahrbericht-tazzari-zero-1124108.html http://www.auto-motor-und-sport.de/fahrberichte/e-auto-tazzari-zero-im-test-e-auto-mit-hohem-gefahrenpotenzial-2785833.html Citroën C1 http://www.sueddeutsche.de/auto/fahrbericht-citroen-c-style-der-preis-ist-heiss-1.552494

Renault Zoe 2012 http://www.goingelectric.de/elektroautos/renault-zoe/ excepted (no price with battery)

German E-Cars CETOS 2013 http://www.green-motors.de/auto/german-e-cars-cetos Opel Corsa http://www.opel-infos.de/infomaterial/20_13_2013-06-24_1.pdf

Citroën C-Zero 2013 http://www.focus.de/auto/elektroauto/citroen-c-zero-kleine-rappelkiste-citroens-mini-stromer_id_3729694.html Citroën C1 http://box.motorline.cc/autowelt/pdf/C1_Produktblatt_2013-03.pdf

Smart ED 2013 http://www.zeit.de/auto/2013-05/smart-elektroauto Smart fortwo https://www.swmb.de/fileadmin/resources/documents/smart%20fortwo%20Preisliste%202013.pdf

VW e-up 2013 http://www.autobild.de/artikel/vw-e-up-iaa-2013-neue-infos-4276629.html VW up! (cheer up) file://cheer-up_preisliste.pdf

Renault Fluence Z.E. 2013 file://Preisliste_Fluence_ZE.pdf excepted (no price with battery)

Ford Focus Electric 2013 http://www.autobild.de/artikel/ford-focus-electric-preise-und-technische-daten-4256882.html Ford Focus 1.0 EcoBoost http://www.autohaus-ristow.de/downloads.html?file=files/ristow/uploads/pdf%20Preislisten/Ford_Focus_Preisliste.pdf

Peugeot iOn 2013 http://www.autobild.de/artikel/peugeot-ion-preis-2773941.html Citroën C1 http://box.motorline.cc/autowelt/pdf/C1_Produktblatt_2013-03.pdf

BMW i3 60 Ah 2013 http://www.goingelectric.de/2013/07/30/news/bmw-i3-preisliste-sonderausstattung/ BMW 320i file://Preisliste_BMW_3er_Limousine_M%C3%A4rz_2013%20(1).pdf

Mitsubishi i-MiEV 2013 Mitsubishi Österreich: Produktseite des i-MiEV (Memento vom 4. Januar 2014 im Internet Archive) // MITSUBISHI MOTORS Deutschland Website des i-MiEV Citroën C1 http://box.motorline.cc/autowelt/pdf/C1_Produktblatt_2013-03.pdf

Renault Kangoo Z.E 2013 http://www.elektroauto-news.net/elektroautos/renault/renault-kangoo-ze-elektro-transporter-fuer-die-city excepted (no price with battery)

Fiat Karabag 500 E 2013 http://www.green-motors.de/auto/karabag-500-e Fiat 500 http://www.fiat-press.de/download/DE/2013/FIAT/Preislisten/130701_F_500_PL.pdf

Nissan Leaf Visia 2013 http://www.auto.de/magazin/vergleich-bmw-i3-gegen-nissan-leaf-es-hat-sich-was-getan/ http://www.goingelectric.de/elektroautos/nissan-leaf-2013/ Ford Focus 1.0 EcoBoost http://www.autohaus-ristow.de/downloads.html?file=files/ristow/uploads/pdf%20Preislisten/Ford_Focus_Preisliste.pdf

Tesla Model S 2013 http://www.autobild.de/artikel/tesla-model-s-preise-in-deutschland-3761908.html BMW 335i https://www.press.bmwgroup.com/deutschland/article/detail/T0136158DE/die-preise-der-bmw-3er-limousine-gueltig-ab-maerz-2013?language=de

Tesla Model S 60 2013 http://www.auto-motor-und-sport.de/news/tesla-model-s-preise-elektro-limousine-startet-bei-71400-euro-6872521.html BMW 335i https://www.press.bmwgroup.com/deutschland/article/detail/T0136158DE/die-preise-der-bmw-3er-limousine-gueltig-ab-maerz-2013?language=de

German E-Cars Stromos 2013	http://www.elektroautomobile.org/app/download/6343536056/Kurzpreisliste+PKW+2012.pdf?t=1355130875	Opel Agila B / Suzuki Splash	http://www.opel-infos.de/infomaterial/2_13_2012-11-19_1.pdf
Tazzari Zero 2013	http://www.autobild.de/artikel/fahrbericht-tazzari-zero-1124108.html	http://www.auto-motor-und-sport.de/fahrberichte/e-auto-tazzari-zero-im-test-e-auto-mit-hohem-gefahrenpotenzial-2785833.html	Citroën C1 http://box.motorline.cc/autowelt/pdf/C1_Produktblatt_2013-03.pdf
Mercedes-Benz B-Klasse Electric Drive 2014	http://www.autobild.de/artikel/mercedes-b-klasse-electric-drive-preise-5424972.html	Mercedes-Benz B 250	http://info.martin-jacoby.com/mercedes-benz/price-lists/B_Klasse_Sports_Tourer_W246_2014-05-26.pdf
German E-Cars CETOS 2014	http://www.focus.de/auto/elektroauto/cetos-unbekannte-elektro-groesse-opel-unter-strom_id_3729736.html	Opel Corsa	http://www.opel-infos.de/infomaterial/20_13_2014-06-30_1.pdf
Citroën C-Zero 2014	http://www.auto-motor-und-sport.de/news/citroen-senkt-e-auto-preise-c-zero-und-berlingo-electric-werden-billiger-774486.html	Citroën C1	http://www.autobild.de/artikel/citroen-c1-2014-preise-5069710.html
Smart ED 2014	https://www.smart.com/content/dam/smart/DE/PDF/Preisliste_smart_fortwo_electric_drive.pdf	Smart fortwo	http://www.autobild.de/artikel/smart-fortwo-forfour-2014-preise-4421069.html
Volkswagen E-Golf 2014	https://www.adac.de/_ext/itr/tests/Autotest/AT5134_VW_e_Golf/VW_e_Golf.pdf	VW Golf TSI BlueMotion Comfortline	file://s/golf_ams1715_052.pdf
VW e-up 2014	http://www.autobild.de/artikel/vw-e-up-test-4544079.html	VW up! (cheer up)	file://cheer-up_preisliste.pdf
Renault Fluence Z.E. 2014	http://www.focus.de/auto/elektroauto/renault-fluence-z-e-elektro-limousine-renault-fluence_id_3721550.html	excepted (no price with battery)	
Ford Focus Electric 2014	http://www.autobild.de/artikel/ford-focus-electric-preise-und-technische-daten-4256882.html	Ford Focus 1.0 EcoBoost	http://www.autobild.de/artikel/ford-focus-2015-preise-5250782.html
Peugeot iOn 2014	http://www.autobild.de/artikel/peugeot-ion-preis-2773941.html	Citroën C1	http://www.autobild.de/artikel/citroen-c1-2014-preise-5069710.html
BMW i3 60 Ah 2014	http://www.t-online.de/auto/news/id_67085938/bmw-i3-preise-das-kostet-das-elektroauto.html	BMW 320i	file://Preise_BMW_3er_Limousine_M%C3%A4rz_2014.pdf
Mitsubishi i-MiEV 2014	Mitsubishi senkt die Preise für seinen Elektro-Cityflitzer, offizielle Pressemeldung vom 24. März 2014	Citroën C1	http://www.autobild.de/artikel/citroen-c1-2014-preise-5069710.html
Renault Kangoo Z.E 2014	http://www.stadelbauer.de/upload/1295624_Preisliste_Kangoo_ZE.pdf	excepted (no price with battery)	
Fiat Karabag 500 E 2014	http://www.focus.de/auto/elektroauto/karabag-new-500e-fiat-unter-strom-der-lautlose-fraueneroberer_id_3731165.html	Fiat 500	http://auto-pfaff.de/files/preisliste_fiat_500_1.pdf

Nissan Leaf Visia 2014 http://www.auto.de/magazin/vergleich-bmw-i3-gegen-nissan-leaf-es-hat-sich-was-getan/ http://www.goingelectric.de/elektroautos/nissan-leaf-2013/ Ford Focus 1.0 EcoBoost http://www.autobild.de/artikel/ford-focus-2015-preise-5250782.html

Tesla Model S 60 2014 http://www.captain-gadget.de/tesla-senkt-den-einstiegspreis-mit-60-kwh-model-s/ BMW 335i https://www.press.bmwgroup.com/deutschland/article/detail/T0164765DE/die-preise-des-bmw-3er-limousine-touring-und-gran-turismo-gueltig-ab-maerz-2014?language=de

Peugeot Partner Electric L1 2014 http://www.autozeitung.de/auto-neuheiten/peugeot-partner-electric-preis-elektroauto-lieferwagen-bilder Peugeot Partner Kastenwagen http://www.autokiste.de/psg/1403/10964.htm

Kia Soul EV 2014 http://www.sueddeutsche.de/auto/kia-soul-ev-im-fahrbericht-schoen-einfach-und-einfach-gut-1.2155289 Kia Soul file://kia-soul-preisliste.pdf

German E-Cars Stromos 2014 http://www.elektroautomobile.org/app/download/6343536056/Kurzpreisliste+PKW+2012.pdf?t=1355130875 Opel Agila B http://www.opel-infos.de/infomaterial/2_13_2012-11-19_1.pdf

Tazzari Zero 2014 http://www.focus.de/auto/fahrberichte/elektroautos/tazzari-zero-italo-smart-tazzaris-stromer-exot_id_3731237.html Citroën C1 http://www.autobild.de/artikel/citroen-c1-2014-preise-5069710.html

Mercedes-Benz B-Klasse Electric Drive 2015 http://www.autobild.de/artikel/mercedes-b-klasse-electric-drive-preise-5424972.html Mercedes-Benz B 250 http://info.martin-jacoby.com/mercedes-benz/price-lists/B_Klasse_Sports_Tourer_W246_2014-05-26.pdf

CCS Elektromobile City-Stromerle Coupe 2015 http://www.emobilserver.de/service-tools/katalog/elektroautos/c/city-stromerle-coupe.html http://www.goingelectric.de/garage/schnarchladers-CityStromerleCoupe/712/ Aixam Miniauto L7e http://www.goingelectric.de/garage/schnarchladers-CityStromerleCoupe/712/ http://aixampreise.de/wp-content/uploads/2015/08/Minauto-Preise.pdf

Citroën C-Zero 2015 http://www.green-motors.de/auto/citroen-c-zero Citroën C1 http://citroen-de-de.custhelp.com/euf/assets/images/allemagne/citroen/Deutschland/C1/C1_K51201.pdf

Smart ED 2015 http://www.smart.at/content/dam/smart/AT/pdf/downloadcenter/Preisliste-smart_electric_drive_01-2015.pdf Smart fortwo http://www.autobild.de/artikel/smart-fortwo-forfour-2014-preise-4421069.html

Volkswagen E-Golf 2015 http://www.autobild.de/artikel/preise-vw-e-golf-2014--4587096.html VW Golf TSI BlueMotion Comfortline file://golf_ams1715_052.pdf

VW e-up 2015 https://www.welt.de/motor/news/article139432795/Test-VW-E-Up.html VW up! (cheer up) file://cheer-up_preisliste.pdf

Ford Focus Electric 2015 http://www.auto-pieroth.de/Global/Ford/PDF/pkw_preislisten/Preisliste_Ford_Focus_Electric.pdf Ford Focus 1.0 EcoBoost http://www.autobild.de/artikel/ford-focus-2015-preise-5250782.html

Peugeot i0n 2015 http://www.automobile-rosenkranz.de/dokumente/peugeot-ion-preise-und-daten.pdf Citroën C1 http://citroen-de-de.custhelp.com/euf/assets/images/allemagne/citroen/Deutschland/C1/C1_K51201.pdf

BMW i3 60 Ah 2015 https://www.welt.de/motor/article137024263/Im-BMW-i3-fuehlt-man-sich-manchmal-verladen.html BMW 320i file://Preisliste_BMW_3er_Touring_Juli_2015.pdf

Mitsubishi	i-MiEV	2015	Mitsubishi senkt die Preise für seinen Elektro-Cityflitzer, offizielle Pressemeldung vom 24. März 2014 Citroën C1 http://citroen-de-de.custhelp.com/euf/assets/images/allemagne/citroen/Deutschland/C1/C1_K51201.pdf	
Renault Kangoo Z.E		2015	http://www.stadelbauer.de/upload/1295624_Preisliste_Kangoo_ZE.pdf	excepted (no price with battery)
Nissan LEAF ACENTA Focus 1.0 EcoBoost		2015	http://www.autobild.de/artikel/nissan-leaf-neue-batterie-5857129.html http://www.autobild.de/artikel/ford-focus-2015-preise-5250782.html	Ford
Nissan LEAF ACENTA Focus 1.0 EcoBoost		2015	http://www.autobild.de/artikel/nissan-leaf-neue-batterie-5857129.html http://www.autobild.de/artikel/ford-focus-2015-preise-5250782.html	Ford
Nissan Leaf Visia		2015	http://www.auto.de/magazin/vergleich-bmw-i3-gegen-nissan-leaf-es-hat-sich-wasgetan/ http://www.goingelectric.de/elektroautos/nissan-leaf-2013/ Ford Focus 1.0 EcoBoost http://www.autobild.de/artikel/ford-focus-2015-preise-5250782.html	
Aixam Mega e-City city.php Aixam Miniauto L6e		2015	keine genaue Datumsangabe, Annahme: 2015 http://www.emission-zero.de/megahttp://aixampreise.de/wp-content/uploads/2015/08/Minauto-Preise.pdf	
Tesla Model S 60 BMW 335i		2015	http://www.captain-gadget.de/tesla-senkt-den-einstiegspreis-mit-60-kwh-model-s/ https://www.press.bmwgroup.com/deutschland/article/detail/T0200579DE/die-preise-der-bmw-3er-limousine-gueltig-ab-maerz-2015?language=de	
Peugeot Partner Electric L1 2015 lieferwagen-bilder Peugeot Partner Kastenwagen preise-5707438.html			http://www.autozeitung.de/auto-neuheiten/peugeot-partner-electric-preis-elektroautohttp://www.autobild.de/artikel/peugeot-partner-facelift-2015-	
Kia Soul EV 2015 Kia Soul file:// /kia-soul-preisliste.pdf			http://www.auto-motor-und-sport.de/einzeltests/kia-soul-ev-elektro-auto-test-9144036.html	
German E-Cars Stromos 2015 auch-diese-elektroautos-kann-man-kaufen many/nscwebsite/de/01_Vehicles/01_PassengerCars/Karl/downloads/KARL_17-5_PRL-D.pdf			https://www.emobilitaetonline.de/news/produkte-und-dienstleistungen/2124-spezialOpel Karl http://www.opel.de/content/dam/Opel/Europe/ger-	
Renault Twizy 45 Cargo price with battery)		2015	http://www.stadelbauer.de/upload/1295638_Preisliste_Twizy.pdf	excepted (no
Renault Twizy 45 Life price with battery)		2015	http://www.stadelbauer.de/upload/1295638_Preisliste_Twizy.pdf	excepted (no
Renault Twizy Cargo price with battery)		2015	http://www.stadelbauer.de/upload/1295638_Preisliste_Twizy.pdf	excepted (no
Renault Twizy Life price with battery)		2015	http://www.stadelbauer.de/upload/1295638_Preisliste_Twizy.pdf	excepted (no
Tazzari Zero 2015 exot_id_3731237.html roen/Deutschland/C1/C1_K51201.pdf			http://www.focus.de/auto/fahrberichte/elektroautos/tazzari-zero-italo-smart-tazzaris-stromerCitroën C1 http://citroen-de-de.custhelp.com/euf/assets/images/allemagne/cit-	

Renault ZOE Intense 2015 http://www.autohaus-griesel.de/pdf/Preisliste_Zoe.pdf excepted (no price with battery)

Renault ZOE Life 2015 http://www.autohaus-griesel.de/pdf/Preisliste_Zoe.pdf excepted (no price with battery)

Mercedes-Benz B-Klasse Electric Drive 2016 file://Preisliste_B-Klasse_w246_160808.pdf Mercedes-Benz B 250 file://Preisliste_B-Klasse_w246_160808.pdf

Citroën C-Zero 2016 http://www.citroen.de/modelle/citroen/citroen-c-zero/ausstattungsvarianten-vergleichen.html Citroën C1 http://www.citroen.de/modelle/citroen/citroen-c1.html

BYD e6 2016 http://www.heise.de/autos/artikel/Elektroautos-BYD-Fenecon-startet-den-Verkauf-des-e6-3099764.html http://ecomento.tv/2016/02/15/crossover-elektroauto-byd-e6-ab-sofort-in-deutschland-erhaeltlich/ Ford Focus 1.0 EcoBoost file://der-neue-ford-focus-preisliste.pdf

Smart ED 2016 http://www.focus.de/auto/elektroauto/fahrbericht-smart-fortwo-electric-drive-er-kostet-nur-noch-17-994-euro-neuer-elektro-smart-im-ersten-test_id_6216207.html Smart fortwo https://www.smart.com/content/dam/smart/DE/PDF/Preisliste-smart-forfour-Maerz2016.pdf

Volkswagen E-Golf 2016 http://www.stromschnell.de/kosten/kostenvergleich-vw-golf-10-tsi-versus-vw-egolf_5121362_5093786.html VW Golf TSI BlueMotion Comfortline file://golf_ams1715_052.pdf

Nissan E-NV200 EVALIA 2016 https://www.nissan.de/fahrzeuge/neuwagen/e-nv200-evalia.html https://www.nissan-cdn.net/content/dam/Nissan/de/brochures/pkw/20160915-nissan-e-nv200-evalia-broschuere-preisliste-de.pdf Nissan EVALIA https://www.nissan.de/fahrzeuge/neuwagen/nv200-evalia/varianten-preise.html

VW e-up 2016 http://www.autobild.de/artikel/vw-e-up-iaa-2013-neue-infos-4276629.html file://e-up_preisliste%20.pdf VW up! (move up!) file:// /up_preisliste%20.pdf

Ford Focus Electric 2016 http://www.ford.de/cs/BlobServer?blobtable=MungoBlobs&blobcol=urldata&blobheadervalue1=attachment%3Bfilename%3D%22Ford+Focus+Electric+-+Preisliste.pdf%22&blobheadervalue2=abinary%3Bcharset%3DUTF-8&blobheadername1=Content-Disposition&blobheadername2=MDT-Type&blobheader=application%2Fpdf&blobwhere=1214487735435&blobkey=id Ford Focus 1.0 EcoBoost file://der-neue-ford-focus-preisliste.pdf

Peugeot i0n 2016 http://media.peugeot.de/file/40/7/preisliste-ion-14.09.2016.95407.pdf#_ga=1.11026679.2126958622.1480018350 Citroën C1 http://www.citroen.de/modelle/citroen/citroen-c1.html

BMW i3 60 Ah 2016 http://www.bmw.de/dam/brandBM/marketDE/countryDE/newvehicles/allfacts/pricelist/BMW_i3_Preisliste.pdf?download.1431068004689.pdf BMW 320i file://BMW_3er_Limousine_Preisliste%20.pdf

BMW i3 94 Ah 2016 http://www.bmw.de/dam/brandBM/marketDE/countryDE/newvehicles/allfacts/pricelist/BMW_i3_Preisliste.pdf?download.1431068004689.pdf BMW 320i file://Preise_BMW_3er_Limousine_M%C3%A4rz_2015.pdf

Mitsubishi i-MiEV 2016 Mitsubishi senkt die Preise für seinen Elektro-Cityflitzer, offizielle Pressemeldung vom 24. März 2014 Citroën C1 http://www.citroen.de/modelle/citroen/citroen-c1.html

| Nissan LEAF ACENTA | 2016 | https://www.nissan.de/fahrzeuge/neuwagen/leaf/varianten-preise.html | Ford Focus 1.0 EcoBoost | file:// /der-neue-ford-focus-preisliste.pdf |

| Nissan LEAF ACENTA | 2016 | https://www.nissan.de/fahrzeuge/neuwagen/leaf/varianten-preise.html | Ford Focus 1.0 EcoBoost | file:// /der-neue-ford-focus-preisliste.pdf |

| Nissan LEAF TEKNA | 2016 | https://www.nissan.de/fahrzeuge/neuwagen/leaf/varianten-preise.html | Ford Focus 1.0 EcoBoost | file:// /der-neue-ford-focus-preisliste.pdf |

| Nissan LEAF TEKNA | 2016 | https://www.nissan.de/fahrzeuge/neuwagen/leaf/varianten-preise.html | Ford Focus 1.0 EcoBoost | file:// /der-neue-ford-focus-preisliste.pdf |

Nissan Leaf Visia 2016 https://www.nissan.de/fahrzeuge/neuwagen/leaf/varianten-preise.html https://www.nissan.de/fahrzeuge/neuwagen/leaf.html Ford Focus 1.0 EcoBoost file:// /der-neue-ford-focus-preisliste.pdf

Nissan Leaf Visia 2016 https://www.nissan-cdn.net/content/dam/Nissan/de/brochures/pkw/20161102-nissan-leaf-broschuere-preisliste-de.pdf https://www.nissan.de/fahrzeuge/neuwagen/leaf/varianten-preise.html Ford Focus 1.0 EcoBoost file://der-neue-ford-focus-preisliste.pdf

Aixam Mega e-City 2016 http://www.leicht-kfz-schippers.de/e-city.htm Aixam City http://aixam.de/city-pack-2/

Tesla Model 90D 2016 http://www.autobild.de/artikel/tesla-model-s-2016-facelift-8897847.html Mercedes-AMG C 63 http://www.mercedes-benz.de/content/media_library/germany/mpc_germany/de/mercedes-benz_deutschland/pkw_emb_nextgen/neufahrzeuge/c-klasse/preislisten_pdf/pkw_c-klasse_limousine1.object-Single-MEDIA.tmp/W_205_C-Lim_PL_2016_ONLINE_211016.pdf

Tesla Model P90D 2016 http://www.autobild.de/artikel/tesla-model-s-2016-facelift-8897847.html Mercedes-AMG C 63 http://www.mercedes-benz.de/content/media_library/germany/mpc_germany/de/mercedes-benz_deutschland/pkw_emb_nextgen/neufahrzeuge/c-klasse/preislisten_pdf/pkw_c-klasse_limousine1.object-Single-MEDIA.tmp/W_205_C-Lim_PL_2016_ONLINE_211016.pdf

Tesla Model S 60 2016 http://www.autobild.de/artikel/tesla-model-s-2016-facelift-8897847.html BMW 340i file://BMW_3er_Limousine_Preisliste%20.pdf

Tesla Model S 60 2016 http://www.autobild.de/artikel/tesla-model-s-2016-facelift-8897847.html BMW 340i file://BMW_3er_Limousine_Preisliste%20.pdf

Tesla Model X 75D 2016 https://www.tesla.com/de_DE/modelx Audi SQ7 https://www.audi.de/dam/nemo/models/misc/pdf/my-2017/preisliste/preisliste_q7_q7-e-tron-quattro_sq7-tdi.pdf

Tesla Model X 90D 2016 https://www.tesla.com/de_DE/modelx Audi SQ7 https://www.audi.de/dam/nemo/models/misc/pdf/my-2017/preisliste/preisliste_q7_q7-e-tron-quattro_sq7-tdi.pdf

Tesla Model X 100D 2016 https://www.tesla.com/de_DE/modelx Audi SQ7 https://www.audi.de/dam/nemo/models/misc/pdf/my-2017/preisliste/preisliste_q7_q7-e-tron-quattro_sq7-tdi.pdf

Peugeot Partner Electric L1 2016 http://www.peugeot-professional.de/prospekte-partner-electric Peugeot Partner Kastenwagen file://WEB_Peugeot_Partner_Kawa_Preisliste_160819.pdf

Kia	Soul EV	2016	http://www.sueddeutsche.de/auto/kia-soul-ev-im-fahrbericht-schoen-einfach-und-einfach-gut-1.2155289	Kia Soul	file://kia-soul-preisliste.pdf
German E-Cars	Stromos	2016	https://www.e-deutschland.de/alle-fahrzeuge/alle-e-autos/e-autos/1768/german-e-cars-stromos	Opel Karl	http://www.autobild.de/artikel/opel-karl-2015-preise-5310665.html
Renault	Twizy 45 Cargo	2016	Broschüre (Email)	excepted (no price with battery)	
Renault	Twizy 45 Life	2016	Broschüre (Email)	excepted (no price with battery)	
Renault	Twizy Cargo	2016	Broschüre (Email)	excepted (no price with battery)	
Renault	Twizy Life	2016	Broschüre (Email)	excepted (no price with battery)	
Tazzari	Zero City	2016	http://www.tazzari-zero.com/de/konfigurator/configure/11421/	Citroën C1	http://www.citroen.de/modelle/citroen/citroen-c1.html
Tazzari	Zero EM1	2016	http://www.tazzari-zero.com/de/konfigurator/configure/10514/	Citroën C1	http://www.citroen.de/modelle/citroen/citroen-c1.html
Tazzari	Zero EM2 Space	2016	http://www.tazzari-zero.com/de/konfigurator/configure/11412/	Citroën C1	http://www.citroen.de/modelle/citroen/citroen-c1.html
Tazzari	Zero Junior	2016	http://www.tazzari-zero.com/de/konfigurator/configure/11431/	Citroën C1	http://www.citroen.de/modelle/citroen/citroen-c1.html
Hyundai	Hyundai IONIQ Elektro	2016	http://www.hyundai.de/Modelle/IONIQ-Elektro.html	Toyota Auris	https://www.toyota.de/automobile/auris/index/prices
Renault	ZOE Intense	2016	http://ecomento.tv/2016/09/30/renault-zoe-mit-400-kilometer-reichweite-kostet-ab-24-900-euro/	excepted (no price with battery)	
Renault	ZOE Life	2016	http://www.autohaus-griesel.de/pdf/Preisliste_Zoe.pdf	excepted (no price with battery)	

List 2: BEVs and equivalent CVs – energy consumption and emission data

Mitsubishi	i-MiEV	2010	http://www.mitsubishi-motors.de/Electric-Vehicle/	Citroën C1	http://citroen-de-de.custhelp.com/euf/assets/images/allemagne/citroen/files_download_center/C1/RderReifen/AltePreislisten/C1_Selection_K2_07_13.pdf
Fiat	Karabag 500 E	2010	http://www.autobild.de/artikel/fiat-500-1.2-karabag-500e-test-2334931.html	Fiat 500	http://www.fiatpress.de/download/DE/2013/FIAT/Preislisten/130701_F_500_PL.pdf
German E-Cars	Stromos	2010	https://www.german-e-cars.de/fahrzeuge/personenwagen/stromos/	Opel Agila B / Suzuki Splash	2010 http://www.opel-infos.de/infomaterial/2_13_2010-04-26_1.pdf
Tazzari	Zero	2010	http://www.focus.de/auto/fahrberichte/elektroautos/tazzari-zero-italo-smart-tazzaris-stromer-exot_id_3731237.html	Citroën C1	http://citroen-de-de.custhelp.com/euf/assets/images/allemagne/citroen/files_download_center/C1/RderReifen/AltePreislisten/C1_Selection_K2_07_13.pdf

| Mitsubishi | i-MiEV | 2011 | http://www.mitsubishi-motors.de/Electric-Vehicle/ | Citroën C1 |

http://citroen-de-de.custhelp.com/euf/assets/images/allemagne/citroen/files_download_center/C1/RderReifen/AltePreislisten/C1_Selection_K2_07_13.pdf

| Fiat Karabag 500 E | 2011 | http://www.autobild.de/artikel/fiat-500-1.2-karabag-500e-test-2334931.html | Fiat 500 | http://www.fiatpress.de/download/DE/2013/FIAT/Preislisten/130701_F_500_PL.pdf |

| Aixam | Mega e-City | 2011 | http://www.emission-zero.de/megacity.php | Aixam / Ligier | Autoscout24 |

| German E-Cars | Stromos | 2011 | https://www.german-e-cars.de/fahrzeuge/personenwagen/stromos/ | Opel Agila B / Suzuki Splash | http://www.opel-infos.de/infomaterial/2_13_2011-11-28_1.pdf |

Tazzari Zero 2011 http://www.focus.de/auto/fahrberichte/elektroautos/tazzari-zero-italo-smart-tazzaris-stromer-exot_id_3731237.html Citroën C1 http://citroen-de-de.custhelp.com/euf/assets/images/allemagne/citroen/files_download_center/C1/RderReifen/AltePreislisten/C1_Selection_K2_07_13.pdf

German E-Cars CETOS 2012 https://www.german-e-cars.de/fahrzeuge/personenwagen/cetos/ Opel Corsa https://de.wikipedia.org/wiki/Opel_Corsa_D

Citroën C-Zero 2012 http://www.citroen.de/modelle/citroen/citroen-c-zero.html Citroën C1 http://citroen-de-de.custhelp.com/euf/assets/images/allemagne/citroen/files_download_center/C1/RderReifen/AltePreislisten/C1_Selection_K2_07_13.pdf

Smart ED 2012 https://www.smart.com/content/dam/smart/DE/PDF/Preisliste_smart_fortwo_electric_drive.pdf Smart fortwo https://www.smart.com/de/de/index/smart-fortwo-453/technical-data.html?s_kwcid=AL!3888!3!124072733729!e!!g!!smart%20fortwo%20verbrauch&ef_id=WIieowAABRkVQB5o:20170125192008:s#engine1 https://de.wikipedia.org/wiki/Smart_Fortwo

Peugeot iOn 2012 http://www2.peugeot.de/shared/ppdownload/download/kataloge/ion.pdf Citroën C1 http://citroen-de-de.custhelp.com/euf/assets/images/allemagne/citroen/files_download_center/C1/RderReifen/AltePreislisten/C1_Selection_K2_07_13.pdf

Mitsubishi i-MiEV 2012 http://www.mitsubishi-motors.de/Electric-Vehicle/ Citroën C1 http://citroen-de-de.custhelp.com/euf/assets/images/allemagne/citroen/files_download_center/C1/RderReifen/AltePreislisten/C1_Selection_K2_07_13.pdf

Fiat Karabag 500 E 2012 http://www.autobild.de/artikel/fiat-500-1.2-karabag-500e-test-2334931.html Fiat 500 http://www.fiatpress.de/download/DE/2013/FIAT/Preislisten/130701_F_500_PL.pdf

Nissan Leaf Visia 2012 https://www.nissan.de/fahrzeuge/neuwagen/leaf.html Ford Focus 1.0 EcoBoost file://der-neue-ford-focus-preisliste.pdf

German E-Cars Stromos 2012 https://www.german-e-cars.de/fahrzeuge/personenwagen/stromos/ Opel Agila B / Suzuki Splash http://www.opel-infos.de/infomaterial/2_13_2012-11-19_1.pdf

Tazzari Zero 2012 http://www.focus.de/auto/fahrberichte/elektroautos/tazzari-zero-italo-smart-tazzaris-stromer-exot_id_3731237.html Citroën C1 http://citroen-de-de.custhelp.com/euf/assets/images/allemagne/citroen/files_download_center/C1/RderReifen/AltePreislisten/C1_Selection_K2_07_13.pdf

German E-Cars   CETOS   2013   https://www.german-e-cars.de/fahrzeuge/personenwagen/cetos/   Opel Corsa   https://de.wikipedia.org/wiki/Opel_Corsa_D

Citroën   C-Zero   2013   http://www.citroen.de/modelle/citroen/citroen-c-zero.html   Citroën C1   http://citroen-de-de.custhelp.com/euf/assets/images/allemagne/citroen/files_download_center/C1/RderReifen/AltePreislisten/C1_Selection_K2_07_13.pdf

Smart   ED   2013   https://www.smart.com/content/dam/smart/DE/PDF/Preisliste_smart_fortwo_electric_drive.pdf   Smart fortwo   https://www.smart.com/de/index/smart-fortwo-453/technical-data.html?s_kwcid=AL!3888!3!124072733729!e!!g!!smart%20fortwo%20verbrauch&ef_id=WIieo-wAABRkVQB5o:20170125192008:s#engine1   https://de.wikipedia.org/wiki/Smart_Fortwo

VW   e-up   2013   file:///e-up_katalog.pdf   VW up! (cheer up)   file://cheer-up_preisliste.pdf

Ford   Focus Electric   2013   http://www.auto-pieroth.de/Global/Ford/PDF/pkw_preislisten/Preisliste_Ford_Focus_Electric.pdf   Ford Focus 1.0 EcoBoost   file://der-neue-ford-focus-preisliste.pdf

Peugeot   iOn   2013   http://www2.peugeot.de/shared/ppdownload/download/kataloge/ion.pdf   Citroën C1   http://citroen-de-de.custhelp.com/euf/assets/images/allemagne/citroen/files_download_center/C1/RderReifen/AltePreislisten/C1_Selection_K2_07_13.pdf

BMW   i3 60 Ah   2013   http://m.bmw.de/content/dam/bmw/marketDE/m_bmw_de/new-vehicles/pdf/BMW_i3_Katalog.pdf.asset.1480437226874.pdf   BMW 320i   file://Preisliste_BMW_3er_Limousine_M%C3%A4rz_2013%20.pdf

Mitsubishi   i-MiEV   2013   http://www.mitsubishi-motors.de/Electric-Vehicle/   Citroën C1   http://citroen-de-de.custhelp.com/euf/assets/images/allemagne/citroen/files_download_center/C1/RderReifen/AltePreislisten/C1_Selection_K2_07_13.pdf

Fiat   Karabag 500 E   2013   http://www.autobild.de/artikel/fiat-500-1.2-karabag-500e-test-2334931.html   Fiat 500   http://www.fiatpress.de/download/DE/2013/FIAT/Preislisten/130701_F_500_PL.pdf

Nissan   Leaf Visia   2013   https://www.nissan.de/fahrzeuge/neuwagen/leaf.html   Ford Focus 1.0 EcoBoost   file://der-neue-ford-focus-preisliste.pdf

Tesla   Model S   2013   https://www.adac.de/_ext/itr/tests/Autotest/AT5022_Tesla_Model_S_Performance/Tesla_Model_S_Performance.pdf   BMW 335i   file://Preisliste_BMW_3er_Limousine_M%C3%A4rz_2013%20(3).pdf

Tesla   Model S 60   2013   https://www.adac.de/_ext/itr/tests/Autotest/AT5022_Tesla_Model_S_Performance/Tesla_Model_S_Performance.pdf   BMW 335i   file://Preisliste_BMW_3er_Limousine_M%C3%A4rz_2013%20(3).pdf

German E-Cars   Stromos   2013   https://www.german-e-cars.de/fahrzeuge/personenwagen/stromos/   Opel Agila B / Suzuki Splash   http://www.opel-infos.de/infomaterial/2_13_2012-11-19_1.pdf

Tazzari   Zero   2013   http://www.focus.de/auto/fahrberichte/elektroautos/tazzari-zero-italo-smart-tazzaris-stromer-exot_id_3731237.html   Citroën C1   http://citroen-de-de.custhelp.com/euf/assets/images/allemagne/citroen/files_download_center/C1/RderReifen/AltePreislisten/C1_Selection_K2_07_13.pdf

Mercedes-Benz   B-Klasse Electric Drive   2014   http://www.mercedes-benz.de/content/germany/mpc/mpc_germany_website/de/home_mpc/passengercars/home/new_cars/models/b-class/w242/facts/drivetrainelectric/drivetrainelectric.html   Mercedes-Benz B 250   https://www.mercedes-benz.de/content/media_library/germany/mpc_germany/de/mercedes-benz_deutschland/pkw_emb_nextgen/neufahrzeuge/b-klasse/preislisten_pdf/pkw_b-klasse_preisliste0.object-SingleMEDIA.download.tmp/Preisliste_B-Klasse_w246_160808.pdf

German E-Cars   CETOS   2014   https://www.german-e-cars.de/fahrzeuge/personenwagen/cetos/   Opel Corsa   https://de.wikipedia.org/wiki/Opel_Corsa_E

Citroën   C-Zero   2014   http://www.citroen.de/modelle/citroen/citroen-c-zero.html   Citroën C1   http://citroen-de-de.custhelp.com/euf/assets/images/allemagne/citroen/Deutschland/C1/C1_K51201.pdf

Smart   ED   2014   https://www.smart.com/content/dam/smart/DE/PDF/Preisliste_smart_fortwo_electric_drive.pdf   Smart fortwo   https://www.smart.com/de/de/index/smart-fortwo-453/technical-data.html?s_kwcid=AL!3888!3!124072733729!e!!g!!smart%20fortwo%20verbrauch&ef_id=WIieowAABRkVQB5o:20170125192008:s#engine1   https://de.wikipedia.org/wiki/Smart_Fortwo

Volkswagen   E-Golf   2014   file://e-golf_preisliste%20.pdf   VW Golf TSI BlueMotion Comfortline   file:///golf_ams1715_052.pdf

VW   e-up   2014   file://e-up_katalog.pdf   VW up! (cheer up)   file://cheer-up_preisliste.pdf

Ford   Focus Electric   2014   http://www.auto-pieroth.de/Global/Ford/PDF/pkw_preislisten/Preisliste_Ford_Focus_Electric.pdf   Ford Focus 1.0 EcoBoost   file://der-neue-ford-focus-preisliste.pdf

Peugeot   i0n   2014   http://www2.peugeot.de/shared/ppdownload/download/kataloge/ion.pdf   Citroën C1   http://citroen-de-de.custhelp.com/euf/assets/images/allemagne/citroen/Deutschland/C1/C1_K51201.pdf

BMW   i3 60 Ah   2014   http://m.bmw.de/content/dam/bmw/marketDE/m_bmw_de/new-vehicles/pdf/BMW_i3_Katalog.pdf.asset.1480437226874.pdf   BMW 320i   file://Preise_BMW_3er_Limousine_M%C3%A4rz_2014.pdf

Mitsubishi   i-MiEV   2014   http://www.mitsubishi-motors.de/Electric-Vehicle/   Citroën C1   http://citroen-de-de.custhelp.com/euf/assets/images/allemagne/citroen/Deutschland/C1/C1_K51201.pdf

Fiat   Karabag 500 E   2014   http://www.autobild.de/artikel/fiat-500-1.2-karabag-500e-test-2334931.html   Fiat 500   http://www.fiatpress.de/download/DE/2013/FIAT/Preislisten/130701_F_500_PL.pdf

Nissan   Leaf Visia   2014   https://www.nissan.de/fahrzeuge/neuwagen/leaf.html   Ford Focus 1.0 EcoBoost   file://der-neue-ford-focus-preisliste.pdf

Tesla   Model S 60   2014   https://www.adac.de/_ext/itr/tests/Autotest/AT5022_Tesla_Model_S_Performance/Tesla_Model_S_Performance.pdf   BMW 335i   file://Preise_BMW_3er_Limousine_M%C3%A4rz_2014%20(5).pdf

Peugeot   Partner Electric L1   2014   file://Partner%20Electric%2008.08.2016%20(7).pdf   Peugeot Partner Kastenwagen   file://WEB_Peugeot_Partner_Kawa_Preisliste_160819.pdf

Kia   Soul EV   2014   http://www.autobild.de/artikel/bmw-i3-kia-soul-ev-test-5394038.html   Kia Soul   file://kia-soul-preisliste.pdf

German E-Cars	Stromos	2014	https://www.german-e-cars.de/fahrzeuge/personenwagen/stromos/		Opel Agila B
	http://www.opel-infos.de/infomaterial/2_13_2012-11-19_1.pdf				
Tazzari Zero		2014	http://www.focus.de/auto/fahrberichte/elektroautos/tazzari-zero-italo-smart-tazzaris-stromer-exot_id_3731237.html		Citroën C1	http://citroen-de-de.custhelp.com/euf/assets/images/allemagne/citroen/Deutschland/C1/C1_K51201.pdf
Mercedes-Benz	B-Klasse Electric Drive	2015	http://www.mercedes-benz.de/content/germany/mpc/mpc_germany_website/de/home_mpc/passengercars/home/new_cars/models/b-class/w242/facts/drivetrainelectric/drivetrainelectric.html Mercedes-Benz B 250	https://www.mercedes-benz.de/content/media_library/germany/mpc_germany/de/mercedes-benz_deutschland/pkw_emb_nextgen/neufahrzeuge/b-klasse/preislisten_pdf/pkw_b-klasse_preisliste0.object-SingleMEDIA.download.tmp/Preisliste_B-Klasse_w246_160808.pdf	
CCS Elektromobile City-Stromerle Coupe		2015	http://www.goingelectric.de/garage/schnarchladers-CityStromerleCoupe/712/ Aixam Miniauto L7e	http://aixampreise.de/wp-content/uploads/2015/08/Minauto-Preise.pdf	
Citroën C-Zero	2015	http://www.citroen.de/modelle/citroen/citroen-c-zero.html Citroën C1			http://citroen-de-de.custhelp.com/euf/assets/images/allemagne/citroen/Deutschland/C1/C1_K51201.pdf
Smart	ED	2015	https://www.smart.com/content/dam/smart/DE/PDF/Preisliste_smart_fortwo_electric_drive.pdf Smart fortwo https://www.smart.com/de/de/index/smart-fortwo-453/technical-data.html?s_kwcid=AL!3888!3!124072733729!e!!g!!smart%20fortwo%20verbrauch&ef_id=WIieowAABRkVQB5o:20170125192008:s#engine1 https://de.wikipedia.org/wiki/Smart_Fortwo		
Volkswagen	E-Golf	2015	file://e-golf_preisliste%20(9).pdf	VW Golf TSI BlueMotion Comfortline file://golf_ams1715_052.pdf	
VW	e-up	2015	file://e-up_katalog.pdf	VW up! (cheer up) file://cheer-up_preisliste.pdf	
Ford	Focus Electric	2015	http://www.auto-pieroth.de/Global/Ford/PDF/pkw_preislisten/Preisliste_Ford_Focus_Electric.pdf Ford Focus 1.0 EcoBoost	file://der-neue-ford-focus-preisliste.pdf	
Peugeot	iOn	2015	http://www2.peugeot.de/shared/ppdownload/download/kataloge/ion.pdf	Citroën C1	
	http://citroen-de-de.custhelp.com/euf/assets/images/allemagne/citroen/Deutschland/C1/C1_K51201.pdf				
BMW	i3 60 Ah	2015	http://m.bmw.de/content/dam/bmw/marketDE/m_bmw_de/new-vehicles/pdf/BMW_i3_Katalog.pdf.asset.1480437226874.pdf BMW 320i	file://Preise_BMW_3er_Limousine_M%C3%A4rz_2015%20.pdf	
Mitsubishi	i-MiEV	2015	http://www.mitsubishi-motors.de/Electric-Vehicle/	Citroën C1	
	http://citroen-de-de.custhelp.com/euf/assets/images/allemagne/citroen/Deutschland/C1/C1_K51201.pdf				
Nissan	LEAF ACENTA	2015	https://www.nissan.de/fahrzeuge/neuwagen/leaf.html	Ford Focus 1.0 EcoBoost file://der-neue-ford-focus-preisliste.pdf	
Nissan	LEAF ACENTA	2015	https://www.nissan.de/fahrzeuge/neuwagen/leaf.html	Ford Focus 1.0 EcoBoost file://der-neue-ford-focus-preisliste.pdf	
Nissan	Leaf Visia	2015	https://www.nissan.de/fahrzeuge/neuwagen/leaf.html	Ford Focus 1.0 EcoBoost file://der-neue-ford-focus-preisliste.pdf	

Aixam	Mega e-City	2015	http://www.emission-zero.de/megacity.php		Aixam Miniauto L6e

http://aixampreise.de/wp-content/uploads/2015/08/Minauto-Preise.pdf

Tesla Model S 60 2015 https://www.adac.de/_ext/itr/tests/Autotest/AT5022_Tesla_Model_S_Performance/Tesla_Model_S_Performance.pdf BMW 335i file://Preise_BMW_3er_Limousine_M%C3%A4rz_2015%20(3).pdf

Peugeot Partner Electric L1 2015 file://Partner%20Electric%2008.08.2016%20(7).pdf Peugeot Partner Kastenwagen file://WEB_Peugeot_Partner_Kawa_Preisliste_160819.pdf

Kia Soul EV 2015 http://www.autobild.de/artikel/bmw-i3-kia-soul-ev-test-5394038.html Kia Soul file://kia-soul-preisliste.pdf

German E-Cars Stromos 2015 https://www.german-e-cars.de/fahrzeuge/personenwagen/stromos/ Opel Karl http://www.opel.de/content/dam/Opel/Europe/germany/nscwebsite/de/01_Vehicles/01_PassengerCars/Karl/downloads/KARL_17-5_PRL-D.pdf

Tazzari Zero 2015 http://www.focus.de/auto/fahrberichte/elektroautos/tazzari-zero-italo-smart-tazzaris-stromer-exot_id_3731237.html Citroën C1 http://citroen-de-de.custhelp.com/euf/assets/images/allemagne/citroen/Deutschland/C1/C1_K51201.pdf

Mercedes-Benz B-Klasse Electric Drive 2016 http://www.mercedes-benz.de/content/germany/mpc/mpc_germany_website/de/home_mpc/passengercars/home/new_cars/models/b-class/w242/facts/drivetrainelectric/drivetrainelectric.html Mercedes-Benz B 250 https://www.mercedes-benz.de/content/media_library/germany/mpc_germany/de/mercedes-benz_deutschland/pkw_emb_nextgen/neufahrzeuge/b-klasse/preislisten_pdf/pkw_b-klasse_preisliste0.object-SingleMEDIA.download.tmp/Preisliste_B-Klasse_w246_160808.pdf

Citroën C-Zero 2016 http://www.citroen.de/modelle/citroen/citroen-c-zero.html Citroën C1 http://citroen-de-de.custhelp.com/euf/assets/images/allemagne/citroen/Deutschland/C1/C1_K51201.pdf

BYD e6 2016 http://byd-elektroauto.de/pdf/BYD-e6-Elektroauto-Prospekt-Trafa-Fenecon.pdf Ford Focus 1.0 EcoBoost file://der-neue-ford-focus-preisliste.pdf

Smart ED 2016 https://www.smart.com/content/dam/smart/DE/PDF/Preisliste_smart_fortwo_electric_drive.pdf Smart fortwo https://www.smart.com/de/de/index/smart-fortwo-453/technical-data.html?s_kwcid=AL!3888!3!124072733729!e!!g!!smart%20fortwo%20verbrauch&ef_id=WIieowAABRkVQB5o:20170125192008:s#engine1 https://de.wikipedia.org/wiki/Smart_Fortwo

Volkswagen E-Golf 2016 file://e-golf_preisliste%20.pdf VW Golf TSI BlueMotion Comfortline file://golf_ams1715_052.pdf

Nissan E-NV200 EVALIA 2016 https://www.adac.de/_ext/itr/tests/Autotest/AT5258_Nissan_e_NV200_Kombi_Premium/Nissan_e_NV200_Kombi_Premium.pdf Nissan EVALIA https://www.nissan.de/fahrzeuge/neuwagen/nv200-evalia/varianten-preise.html

VW e-up 2016 file://e-up_katalog.pdf VW up! (move up!) file://up_preisliste%20(1).pdf

Ford Focus Electric 2016 http://www.auto-pieroth.de/Global/Ford/PDF/pkw_preislisten/Preisliste_Ford_Focus_Electric.pdf Ford Focus 1.0 EcoBoost file://der-neue-ford-focus-preisliste.pdf

| Hyundai | Hyundai IONIQ Elektro | 2016 | http://www.hyundai.de/Modelle/IONIQ-Elektro.html | Toyota Auris file://Toyota-Auris-2015-PDB_tcm-17-484491.pdf |

| Peugeot | i0n | 2016 | http://www2.peugeot.de/shared/ppdownload/download/kataloge/ion.pdf | Citroën C1 http://citroen-de-de.custhelp.com/euf/assets/images/allemagne/citroen/Deutschland/C1/C1_K51201.pdf |

| BMW | i3 60 Ah | 2016 | http://m.bmw.de/content/dam/bmw/marketDE/m_bmw_de/new-vehicles/pdf/BMW_i3_Katalog.pdf.asset.1480437226874.pdf | BMW 320i | file://Preise_BMW_3er_Limousine_M%C3%A4rz_2015%20.pdf |

| BMW | i3 94 Ah | 2016 | http://www.autobild.de/artikel/bmw-i3-2016-neuer-akku-mehr-reichweite-8550999.html | | BMW 320i | http://m.bmw.de/content/dam/bmw/marketDE/m_bmw_de/new-vehicles/pdf/BMW_3er_Limousine_Touring_Katalog.pdf.asset.1480006133200.pdf |

| Mitsubishi | i-MiEV | 2016 | http://www.mitsubishi-motors.de/Electric-Vehicle/ | Citroën C1 http://citroen-de-de.custhelp.com/euf/assets/images/allemagne/citroen/Deutschland/C1/C1_K51201.pdf |

| Nissan | LEAF ACENTA | 2016 | https://www.nissan.de/fahrzeuge/neuwagen/leaf.html | Ford Focus 1.0 EcoBoost file://der-neue-ford-focus-preisliste.pdf |

| Nissan | LEAF ACENTA | 2016 | https://www.nissan.de/fahrzeuge/neuwagen/leaf.html | Ford Focus 1.0 EcoBoost file://der-neue-ford-focus-preisliste.pdf |

| Nissan | LEAF TEKNA | 2016 | https://www.nissan.de/fahrzeuge/neuwagen/leaf.html | Ford Focus 1.0 EcoBoost file://der-neue-ford-focus-preisliste.pdf |

| Nissan | LEAF TEKNA | 2016 | https://www.nissan.de/fahrzeuge/neuwagen/leaf.html | Ford Focus 1.0 EcoBoost file://der-neue-ford-focus-preisliste.pdf |

| Nissan | Leaf Visia | 2016 | https://www.nissan.de/fahrzeuge/neuwagen/leaf.html | Ford Focus 1.0 EcoBoost file://der-neue-ford-focus-preisliste.pdf |

| Nissan | Leaf Visia | 2016 | https://www.nissan.de/fahrzeuge/neuwagen/leaf.html | Ford Focus 1.0 EcoBoost file://der-neue-ford-focus-preisliste.pdf |

| Aixam | Mega e-City | 2016 | http://www.emission-zero.de/megacity.php | Aixam City | http://aixam.de/city-pack-2/ |

Tesla   Model 90D   2016   Performance (P) models have the same energy demand, see http://www.autobild.de/artikel/tesla-model-s-p85d-langstrecken-test-6032597.html   Mercedes-AMG C 63   http://www.mercedes-benz.de/content/media_library/germany/mpc_germany/de/mercedes-benz_deutschland/pkw_emb_nextgen/neufahrzeuge/c-klasse/preislisten_pdf/pkw_c-klasse_limousine1.object-Single-MEDIA.tmp/W_205_C-Lim_PL_2016_ONLINE_211016.pdfhttp://www.mercedes-benz.de/content/media_library/germany/mpc_germany/de/mercedes-benz_deutschland/pkw_emb_nextgen/neufahrzeuge/c-klasse/preislisten_pdf/pkw_c-klasse_limousine1.object-Single-MEDIA.tmp/W_205_C-Lim_PL_2016_ONLINE_211016.pdf

Tesla   Model P90D   2016   Performance (P) models have the same energy demand, see http://www.autobild.de/artikel/tesla-model-s-p85d-langstrecken-test-6032597.html   Mercedes-AMG C 63   http://www.mercedes-benz.de/content/media_library/germany/mpc_germany/de/mercedes-benz_deutschland/pkw_emb_nextgen/neufahrzeuge/c-klasse/preislisten_pdf/pkw_c-klasse_limousine1.object-Single-MEDIA.tmp/W_205_C-

Lim_PL_2016_ONLINE_211016.pdfhttp://www.mercedes-benz.de/content/media_library/germany/mpc_germany/de/mercedes-benz_deutschland/pkw_emb_nextgen/neufahrzeuge/c-klasse/preislisten_pdf/pkw_c-klasse_limousine1.object-SingleMEDIA.tmp/W_205_C-Lim_PL_2016_ONLINE_211016.pdf

Tesla	Model S 60	2016	https://www.adac.de/_ext/itr/tests/Autotest/AT5022_Tesla_Model_S_Performance/Tesla_Model_S_Performance.pdf	
			BMW 340i	file://BMW_3er_Limousine_Preisliste%20.pdf

Tesla　Model S 60　2016　https://www.adac.de/_ext/itr/tests/Autotest/AT5022_Tesla_Model_S_Performance/Tesla_Model_S_Performance.pdf　BMW 340i　file://BMW_3er_Limousine_Preisliste%20.pdf

Tesla　Model X 75D　2016　https://www.adac.de/_ext/itr/tests/Autotest/AT5022_Tesla_Model_S_Performance/Tesla_Model_S_Performance.pdf　Audi SQ7　https://www.audi.de/dam/nemo/models/misc/pdf/my-2017/preislisten/preisliste_q7_q7-e-tron-quattro_sq7-tdi.pdf

Tesla　Model X 90D　2016　https://www.adac.de/_ext/itr/tests/Autotest/AT5022_Tesla_Model_S_Performance/Tesla_Model_S_Performance.pdf　Audi SQ7　https://www.audi.de/dam/nemo/models/misc/pdf/my-2017/preislisten/preisliste_q7_q7-e-tron-quattro_sq7-tdi.pdf

Tesla　Model X 100D　2016　https://www.adac.de/_ext/itr/tests/Autotest/AT5022_Tesla_Model_S_Performance/Tesla_Model_S_Performance.pdf　Audi SQ7　https://www.audi.de/dam/nemo/models/misc/pdf/my-2017/preislisten/preisliste_q7_q7-e-tron-quattro_sq7-tdi.pdf

Peugeot　Partner Electric L1 2016　file://Partner%20Electric%2008.08.2016%20(7).pdf　Peugeot Partner Kastenwagen　file://WEB_Peugeot_Partner_Kawa_Preisliste_160819.pdf

Kia　Soul EV 2016　http://www.autobild.de/artikel/bmw-i3-kia-soul-ev-test-5394038.html　Kia Soul file://kia-soul-preisliste.pdf

German E-Cars　Stromos 2016　https://www.german-e-cars.de/fahrzeuge/personenwagen/stromos/　Opel Karl http://www.opel.de/content/dam/Opel/Europe/germany/nscwebsite/de/01_Vehicles/01_PassengerCars/Karl/downloads/KARL_17-5_PRL-D.pdf

Tazzari　Zero City 2016　http://www.focus.de/auto/fahrberichte/elektroautos/tazzari-zero-italo-smart-tazzaris-stromer-exot_id_3731237.html　Citroën C1　http://citroen-de-de.custhelp.com/euf/assets/images/allemagne/citroen/Deutschland/C1/C1_K51201.pdf

Tazzari　Zero EM1　2016　http://www.focus.de/auto/fahrberichte/elektroautos/tazzari-zero-italo-smart-tazzaris-stromer-exot_id_3731237.html　Citroën C1　http://citroen-de-de.custhelp.com/euf/assets/images/allemagne/citroen/Deutschland/C1/C1_K51201.pdf

Tazzari　Zero EM2 Space 2016　http://www.focus.de/auto/fahrberichte/elektroautos/tazzari-zero-italo-smart-tazzaris-stromer-exot_id_3731237.html　Citroën C1　http://citroen-de-de.custhelp.com/euf/assets/images/allemagne/citroen/Deutschland/C1/C1_K51201.pdf

Tazzari　Zero Junior　2016　http://www.focus.de/auto/fahrberichte/elektroautos/tazzari-zero-italo-smart-tazzaris-stromer-exot_id_3731237.html　Citroën C1　http://citroen-de-de.custhelp.com/euf/assets/images/allemagne/citroen/Deutschland/C1/C1_K51201.pdf

List 3: PHEVs and equivalent CVs – purchasing price data

Opel Ampera 2011 http://ww2.autoscout24.de/erste-infos/opel-ampera/mit-spannung-erwartet/44331/106921/?article=106921 http://www.auto-motor-und-sport.de/fahrberichte/chevrolet-volt-im-fahrbericht-wie-sparsam-ist-das-e-auto-wirklich-3555740.html Opel Astra http://www.opel-infos.de/infomaterial/73_13_2011-11-28_1.pdf

Chevrolet Volt 2011 http://www.zeit.de/auto/2011-10/chevrolet-elektroauto-volt http://www.auto-motor-und-sport.de/fahrberichte/chevrolet-volt-im-fahrbericht-wie-sparsam-ist-das-e-auto-wirklich-3555740.html http://www.n-tv.de/auto/Volt-kostet-41-950-Euro-article2731061.html Opel Astra http://www.opel-infos.de/infomaterial/73_13_2011-11-28_1.pdf

Opel Ampera 2012 Wikipedia -> http://www.opel.de/content/dam/Opel/Europe/germany/nscwebsite/de/01_Vehicles/01_PassengerCars/Ampera/katalog/Ampera_Preisliste_20120801_1.0.pdf Opel Astra http://www.opel-infos.de/infomaterial/73_13_2011-11-28_1.pdf

Fisker Karma 2012 http://www.autobild.de/artikel/fisker-karma-preise-2799446.html BMW 650i Gran Coupe file://Preisliste_BMW_6er_Gran_Coup%C3%A9_Juli_2012%20(1).pdf

Toyota Prius III 2012 http://www.spiegel.de/auto/aktuell/toyota-prius-plug-in-hybrid-das-teure-teilzeit-elektroauto-a-872425.html Toyota Auris http://www.autobild.de/artikel/toyota-auris-preis-3694993.html

Volvo V60 D6 Twin Engine 2012 http://www.auto-motor-und-sport.de/fahrberichte/volvo-v60-plug-in-hybrid-nordische-kombination-mal-anders-4855079.html Volvo V60 D5 Momentum http://www.mvcmotors.at/wp-content/uploads/pdf/volvo/neuwagen/Preisliste_Volvo_V60.pdf

Chevrolet Volt 2012 http://www.green-motors.de/auto/chevrolet-volt Opel Astra http://www.opel-infos.de/infomaterial/73_13_2011-11-28_1.pdf

Porsche 918 Spyder 2013 only 918 built, http://www.spiegel.de/auto/aktuell/weltpremiere-918-spyder-porsche-praesentiert-drei-liter-supersportwagen-a-681135.html excepted (supercar)

Opel Ampera 2013 http://www.opel-infos.de/infomaterial/75_13_2013-09-09_1.pdf Opel Astra http://www.opel-infos.de/infomaterial/73_13_2011-11-28_1.pdf

BMW i3 60 Ah with range extender 2013 http://www.goingelectric.de/2013/07/30/news/bmw-i3-preisliste-sonderausstattung/ BMW 320i file://Preisliste_BMW_3er_Limousine_M%C3%A4rz_2013%20.pdf

Porsche Panamera S E-Hybrid 2013 http://www.autobild.de/artikel/porsche-panamera-plug-in-hybrid-test-4364966.html Porsche Panamera 4 http://www.autozeitung.de/auto-neuheiten/porsche-panamera-facelift-2013-s-e-hybrid-shanghai-motor-show-gn-399667

Toyota Prius III 2013 http://www.spiegel.de/auto/aktuell/toyota-prius-plug-in-hybrid-das-teure-teilzeit-elektroauto-a-872425.html Toyota Auris http://www.autobild.de/artikel/toyota-auris-preis-3694993.html

Volvo V60 D6 Twin Engine 2013 http://www.auto-motor-und-sport.de/einzeltests/volvo-v60-plug-in-hybrid-allrad-diesel-unter-strom-6871919.html Volvo V60 D5 Momentum http://www.mvcmotors.at/wp-content/uploads/pdf/volvo/neuwagen/Preisliste_Volvo_V60.pdf

Porsche 918 Spyder 2014 nur 918 Exempare http://www.spiegel.de/auto/aktuell/weltpremiere-918-spyder-porsche-praesentiert-drei-liter-supersportwagen-a-681135.html excepted (supercar)

Audi A3 e-tron 2014 http://www.focus.de/auto/elektroauto/praxistest-audi-a3-e-tron-kosten-wertverlust-und-fazit_id_4469557.html http://www.auto-motor-und-sport.de/news/audi-a3-sportback-e-tron-auf-der-iaa-plug-in-hybrid-auch-als-fuenftuerer-7694709.html http://www.grueneautos.com/2014/08/verkaufsstart-des-audi-a3-sportback-e-tron-ab-37900-euro/ Audi A3 http://www.audi-partner.de/etc/medialib/ngw/product/pdf/preislisten_my_2014.Par.0043.File.pdf/preisliste_a3-limousine.pdf

Opel Ampera 2014 http://www.opel-infos.de/infomaterial/75_13_2014-06-02_1.pdf Opel Astra http://www.opel-infos.de/infomaterial/73_13_2014-12-15_1.pdf

Porsche Cayenne S E-Hybrid 2014 http://www.autobild.de/artikel/porsche-cayenne-plug-in-hybrid-fahrbericht-5394101.html Porsche Cayenne S http://www.autobild.de/artikel/porsche-cayenne-facelift-2014-preise-5223323.html

VW Golf GTE 2014 http://www.auto-motor-und-sport.de/fahrberichte/vw-golf-gte-hybrid-golf-kann-auch-sportlich-8577199.html VW Golf http://www.autobild.de/artikel/vw-golf-sportsvan-2014-preise-4499018.html

BMW i3 60 Ah with range extender 2014 http://www.t-online.de/auto/news/id_67085938/bmw-i3-preise-das-kostet-das-elektroauto.html BMW 320i file:///Preise_BMW_3er_Limousine_M%C3%A4rz_2014.pdf

BMW i8 2014 http://www.autobild.de/artikel/bmw-i8-preis-5064684.html Audi r8 http://www.spiegel.de/auto/fahrberichte/audi-r8-v10-plus-audis-neuer-supersportler-im-test-a-863065.html

Mitsubishi Outlander PHEV 2014 http://www.focus.de/auto/elektroauto/praxistest-mitsubishi-outlander-plug-in-hybrid-kosten-wertverlust-und-verbrauch_id_4228733.html https://www.adac.de/infotestrat/adac-im-einsatz/motorwelt/mitsubishi_outlander_test.aspx Mitsubishi Outlander 2.0 MIVEC 4WD http://www.autobild.de/artikel/mitsubishi-outlander-phev-preis-gesenkt-5039075.html

Porsche Panamera E-Hybrid 2014 http://www.autobild.de/artikel/porsche-panamera-plug-in-hybrid-test-4364966.html Porsche Panamera 4 http://www.autozeitung.de/auto-neuheiten/porsche-panamera-facelift-2013-s-e-hybrid-shanghai-motor-show-gn-399667

Toyota Prius III 2014 file://Toyota-Prius-Plugin-Hybrid-PDB_tcm-17-100621.pdf Toyota Auris http://www.autobild.de/artikel/toyota-auris-preis-3694993.html

Mercedes-Benz S 500 e 2014 http://www.auto-motor-und-sport.de/news/preis-mercedes-s500-plug-in-hybrid-6-jahre-batterie-garantie-8498641.html Mercedes-Benz S 500 assumption as prices for S500 and S500e are identical in 2015 and 2016 - http://www.auto-motor-und-sport.de/news/preis-mercedes-s500-plug-in-hybrid-6-jahre-batterie-garantie-8498641.html

Volvo V60 D6 Twin Engine 2014 http://www.auto-motor-und-sport.de/einzeltests/volvo-v60-plug-in-hybrid-allrad-diesel-unter-strom-6871919.html Volvo V60 D5 Momentum http://www.mvcmotors.at/wp-content/uploads/pdf/volvo/neuwagen/Preisliste_Volvo_V60.pdf

Porsche 918 Spyder 2015 nur 918 Exempare http://www.spiegel.de/auto/aktuell/weltpremiere-918-spyder-porsche-praesentiert-drei-liter-supersportwagen-a-681135.html excepted (supercar)

Audi A3 e-tron 2015 http://www.focus.de/auto/elektroauto/praxistest-audi-a3-e-tron-kosten-wertverlust-und-fazit_id_4469557.html http://www.auto-motor-und-sport.de/news/audi-a3-sportback-e-tron-auf-der-iaa-plug-in-hybrid-auch-als-fuenftuerer-7694709.html http://www.grueneautos.com/2014/08/verkaufsstart-des-audi-a3-sportback-e-tron-ab-37900-

euro/ Audi A3 https://www.audi.de/dam/nemo/models/misc/pdf/my-2015/preislisten/preisliste_a3_a3-sportback_s3_s3-sportback_20150226.pdf

Mercedes-Benz C 350 e Kombi 2015 http://www.auto-motor-und-sport.de/vergleichstest/vw-passat-gte-mercedes-c-350-e-plug-in-vergleich-10125286.htm Mercedes-Benz C300 Kombi http://www.autobild.de/artikel/mercedes-c-klasse-neue-modelle-alle-preise-5704851.html

Mercedes-Benz C 350 e Limousine 2015 http://www.autozeitung.de/auto-neuheiten/mercedes-c-350-e-preis-plug-in-hybrid http://www.auto-motor-und-sport.de/vergleichstest/vw-passat-gte-mercedes-c-350-e-plug-in-vergleich-10125286.html https://www.mercedes-benz.com/de/mercedes-benz/fahrzeuge/personenwagen/c-klasse/der-neue-c-350-e-effizienz-dynamik-und-komfort/ Mercedes-Benz C300 Limousine http://www.autobild.de/artikel/mercedes-c-klasse-neue-modelle-alle-preise-5704851.html

Porsche Cayenne S E-Hybrid 2015 http://www.faz.net/aktuell/technik-motor/auto-verkehr/fahrberichte/test-fahrt-mit-dem-porsche-cayenne-s-e-hybrid-13457664-p2.html Porsche Cayenne S http://www.autobild.de/artikel/porsche-cayenne-facelift-2014-preise-5223323.html

Mercedes-Benz GLC 350e 2015 http://www.n-tv.de/auto/praxistest/Erster-Praxistest-im-Mercedes-GLC-350e-article17246681.html http://www.auto-motor-und-sport.de/fahrberichte/mercedes-glc-350-e-fahrbericht-10304451.html Mercedes-Benz GLC 350 d 4MATIC http://www.auto-motor-und-sport.de/mercedes/glc/technische-daten/

Mercedes-Benz GLE 500 e 4MATIC 2015 http://www.auto-motor-und-sport.de/news/preise-mercedes-gle-frisch-gemacht-und-umgetauft-ab-53967-euro-689205.html Mercedes-Benz GLE 500 4MATIC http://www.auto-motor-und-sport.de/news/preise-mercedes-gle-frisch-gemacht-und-umgetauft-ab-53967-euro-689205.html

VW Golf GTE 2015 http://www.faz.net/aktuell/technik-motor/auto-verkehr/fahrberichte/fahrbericht-vw-golf-gte-einen-geldwerten-vorteil-hat-nur-das-finanzamt-13636496.html VW Golf http://www.autobild.de/artikel/vw-golf-sportsvan-2014-preise-4499018.html

BMW i3 60 Ah with rage extender 2015 https://www.welt.de/motor/article137024263/Im-BMW-i3-fuehlt-man-sich-manchmal-verladen.html BMW 320i file://Preisliste_BMW_3er_Touring_Juli_2015.pdf

BMW i8 2015 http://www.autobild.de/artikel/bmw-i8-preis-5064684.html Audi r8 http://www.audi-hamburg.de/content/dam/ngw/product/pdf/my-2015/preislisten/preisliste_r8-coupe.pdf

Mitsubishi Outlander PHEV 2015 http://www.focus.de/auto/elektroauto/praxistest-mitsubishi-outlander-plug-in-hybrid-kosten-wertverlust-und-verbrauch_id_4228733.html https://www.adac.de/infotestrat/adac-im-einsatz/motor-welt/mitsubishi_outlander_test.aspx Mitsubishi Outlander 2.0 MIVEC 4WD http://www.autobild.de/artikel/mitsubishi-outlander-phev-preis-gesenkt-5039075.html

Porsche Panamera E-Hybrid 2015 http://www.autobild.de/artikel/porsche-panamera-plug-in-hybrid-test-4364966.html Porsche Panamera 4 http://www.autozeitung.de/auto-neuheiten/porsche-panamera-facelift-2013-s-e-hybrid-shanghai-motor-show-gn-399667

VW Passat GTE 2015 http://www.auto-motor-und-sport.de/vergleichstest/vw-passat-gte-mercedes-c-350-e-plug-in-vergleich-10125286.html VW Passat TDI Bluemotion http://www.focus.de/auto/fahrberichte/fahrbericht-vw-passat-variant-2014-kosten-und-wertverlust_id_4184521.html

VW	Passat Variant GTE	2015	http://www.autobild.de/bilder/vw-passat-gte-2015-preise-5347721.html#bild1	VW Passat Variant TDI Bluemotion	http://www.focus.de/auto/fahrberichte/fahrbericht-vw-passat-variant-2014-kosten-und-wertverlust_id_4184521.html

Toyota	Prius III	2015	file://Toyota-Prius-Plugin-Hybrid-PDB_tcm-17-100621.pdf	Toyota Auris	http://www.autozeitung.de/auto-news/toyota-auris-facelift-2015-preis

Mercedes-Benz	S 500 e	2015	http://www.spiegel.de/auto/fahrberichte/mercedes-s500-plug-in-hybrid-s-klasse-mit-stromanschluss-im-test-a-991857.html	http://www.faz.net/aktuell/technik-motor/auto-verkehr/fahrberichte/fahrbericht-mercedes-s-500-der-lautlose-luxus-13484563.html	http://www.focus.de/auto/elektroauto/praxistest-mercedes-s-500-plug-in-hybrid-steht-ne-s-klasse-vor-dem-aldi_id_4885630.html	Mercedes-Benz S 500	http://www.focus.de/auto/elektroauto/praxistest-mercedes-s-500-plug-in-hybrid-steht-ne-s-klasse-vor-dem-aldi_id_4885630.html

Volvo	V60 D6 Twin Engine	2015	http://www.auto.de/magazin/test-volvo-v60-plug-hybrid-ehrliche-haut/	Volvo V60 D5 Momentum	http://www.mvcmotors.at/wp-content/uploads/pdf/volvo/neuwagen/Preisliste_Volvo_V60.pdf

BMW	X5 xDrive40e	2015	http://www.autobild.de/artikel/bmw-x5-xdrive-40e-2015-fahrbericht-4580148.html	BMW X5 xDrive 35i	file://Preisliste_BMW_X5_April_2014.pdf

Volvo	XC90 T8 Twin Engine	2015	http://www.focus.de/auto/news/volvo-xc90-t8-twin-engine-sparsamer-als-gedacht_id_4649876.html	Volvo XC90 T6 AWD Momentum	http://www.auto-motor-und-sport.de/news/preise-volvo-xc90-schweden-suv-ab-53400-euro-8636897.html

BMW	225xe Active Tourer	2016	http://www.autobild.de/artikel/bmw-225xe-active-tourer-2016-fahrbericht-5840089.html	http://www.meinauto.de/bmw/neuwagen/2er/angebote/2er-active-tourer-plug-in-hybrid	BMW 225d	https://www.bmw.de/dam/brandBM/marketDE/countryDE/newvehicles/allfacts/pricelist/BMW_2er_Coupe_Preisliste.pdf.download.1474024000065.pdf

BMW	330e	2016	file:///BMW_3er_Limousine_Preisliste%20(3).pdf	http://www.autozeitung.de/auto-neuheiten/bmw-330e-2016-preis-technische-daten	BMW 330i	file://BMW_3er_Limousine_Preisliste%20.pdf

BMW	740e	2016	http://www.auto-motor-und-sport.de/news/bmw-740e-iperformance-hybrid-7er-2-1-liter-verbrauch-10524762.html	https://www.press.bmwgroup.com/deutschland/article/detail/T0261925DE/der-neue-bmw-740e-iperformance-der-neue-bmw-740le-iperformance-der-neue-bmw-740le-xdrive-iperformance?language=dehttp://www.sueddeutsche.de/auto/plug-in-hybrid-im-test-der-vierzylinder-ist-der-fehler-im-system-des-bmw-e-1.3106085-2	BMW 740i	file://BMW_7er_Preisliste%20(1).pdf

Audi	A3 e-tron	2016	https://www.audi.de/de/brand/de/neuwagen/a3/a3-sportback-e-tron/motor.html	Audi A3	https://www.audi.de/dam/nemo/models/misc/pdf/my-2015/preislisten/preisliste_a3_a3-sportback_s3_s3-sportback_20150226.pdf

Mercedes-Benz	C 350 e Kombi	2016	file://S_205_C-T_PL_2016_ONLINE_211016.pdf	Mercedes-Benz C300 Kombi	file://S_205_C-T_PL_2016_ONLINE_211016.pdf

Mercedes-Benz	C 350 e Limousine	2016	http://www.mercedes-benz.de/content/media_library/germany/mpc_germany/de/mercedes-benz_deutschland/pkw_emb_nextgen/neufahrzeuge/c-klasse/preislisten_pdf/pkw_c-klasse_limousine1.object-Single-MEDIA.tmp/W_205_C-Lim_PL_2016_ONLINE_211016.pdf	Mercedes-Benz C300 Limousine http://www.mercedes-benz.de/content/media_library/germany/mpc_germany/de/mercedes-benz_deutschland/pkw_emb_nextgen/neufahrzeuge/c-klasse/preislisten_pdf/pkw_c-klasse_limousine1.object-Single-MEDIA.tmp/W_205_C-Lim_PL_2016_ONLINE_211016.pdf

Porsche	Cayenne S E-Hybrid	2016	http://www.porsche.com/germany/models/cayenne/cayenne-s-e-hybrid/
	Porsche Cayenne S		http://www.porsche.com/germany/models/cayenne/cayenne-s/

Mercedes-Benz GLC 350e 2016 file://Preisliste_GLC_SUV_161006.pdf Mercedes-Benz GLC 350 d 4MATIC file://Preisliste_GLC_SUV_161006.pdf

Mercedes-Benz GLE 500 e 4MATIC 2016 file://Preisliste_GLE_166_161006%20.pdf http://www.focus.de/auto/elektroauto/praxistest-mercedes-gle-500e-plug-in-hybrid-dieses-auto-ist-wie-ein-gruener-mit-zuviel-wohlstandsspeck_id_6172608.html Mercedes-Benz GLE 500 4MATIC file://Preisliste_GLE_166_161006%20.pdf

VW Golf GTE 2016 file://golf-gte-preisliste.pdf VW Golf file://golf_preisliste%20.pdf

BMW i3 60 Ah with Rangeextender 2016 http://www.bmw.de/dam/brandBM/marketDE/countryDE/newvehicles/allfacts/pricelist/BMW_i3_Preisliste.pdf?download.1431068004689.pdf BMW 320i file:///BMW_3er_Limousine_Preisliste%20.pdf

BMW i3 94 Ah with Rangeextender 2016 http://www.bmw.de/dam/brandBM/marketDE/countryDE/newvehicles/allfacts/pricelist/BMW_i3_Preisliste.pdf?download.1431068004689.pdf BMW 320i file://Preise_BMW_3er_Limousine_M%C3%A4rz_2015.pdf

BMW i8 2016 https://www.bmw.de/dam/brandBM/marketDE/countryDE/newvehicles/allfacts/pricelist/BMW_i8_Preisliste.pdf.download.1473760565356.pdf Audi r8 https://www.audi.de/dam/nemo/models/misc/pdf/my-2017/preislisten/preisliste_r8-coupe_r8-spyder.pdf

Kia Optima Plug-in-Hybrid 2016 https://www.springerprofessional.de/fahrzeugtechnik/plug-in-hybrid/das-ist-kias-erster-plug-in-hybrid-/10646512 http://www.kia.com/de/modelle/optima-phev/ http://www.autozeitung.de/auto-neuheiten/kia-optima-hybrid-2016-technische-daten-preis Kia Optima http://www.kia.com/de/modelle/optima/

Mitsubishi Outlander PHEV 2016 http://www.focus.de/auto/elektroauto/praxistest-mitsubishi-outlander-plug-in-hybrid-wunder-suv-oder-mogelpackung_id_5685947.html Mitsubishi Outlander 2.0 MIVEC 4WD http://www.mitsubishi-motors.de/workarea/downloadasset.aspx?id=23622332116

Porsche Panamera E-Hybrid 2016 http://www.spiegel.de/auto/fahrberichte/fahrbericht-der-porsche-panameras-e-hybrid-im-test-a-1016664.html Porsche Panamera 4 http://www.porsche.com/germany/models/panamera/panamera-4/

VW Passat GTE 2016 file://passat-gte-passat-gte-variant_preisliste.pdf VW Passat TDI Bluemotion file://passat-passatvariant_preisliste.pdf

VW Passat Variant GTE 2016 file://passat-gte-passat-gte-variant_preisliste.pdf VW Passat Variant TDI Bluemotion file://passat-passatvariant_preisliste.pdf

Toyota Prius IV 2016 http://www.autobild.de/artikel/toyota-prius-2016-vorstellung-5851109.html https://de.wikipedia.org/wiki/Toyota_Prius#Prius_I_.28NHW10.2FNHW11.2C_1997.E2.80.932003.29 Toyota Auris https://www.toyota.de/automobile/auris/index/prices

Audi   Q7 e-tron 2016   http://www.mein-elektroauto.com/fahrzeuge/plug-in-hybridauto-audi-q7-e-tron/   Audi Q7   https://www.audi.de/dam/nemo/models/misc/pdf/my-2015/preislisten/preisliste_q7.pdf

Mercedes-Benz   S 500 e   2016   file:///Preisliste_S-Klasse_Limousine_160812.pdf   Mercedes-Benz S 500   file://Preisliste_S-Klasse_Limousine_160812.pdf

Volvo   V60 D6 Twin Engine   2016   http://www.meinauto.de/testberichte/volvo-v60-d6-twin-engine-im-test-volvos-mittelklasse-kombi-als-plug-in-hybrid   http://www.volvocars.com/de-ch/modelle/unsere-modelle/volvo-v60/editionen/twin-engine   Volvo V60 D5 Momentum   file://Preisliste%20Volvo%20V60%20MY17.pdf

BMW   X5 xDrive40e   2016   file://BMW_X5_Preisliste.pdf   BMW X5 xDrive 35i   file://BMW_X5_Preisliste.pdf

Volvo   XC90 T8 Twin Engine   2016   file://Preisliste%20Volvo%20XC90%20MY17.pdf   Volvo XC90 T6 AWD Momentum   file://Preisliste%20Volvo%20XC90%20MY17.pdf

List 4: PHEVs and equivalent CVs – energy consumption and emission data

Opel   Ampera   2011   http://www.opel-infos.de/infomaterial/75_13_2014-06-02_1.pdf   Opel Astra   http://www.opel-infos.de/infomaterial/73_13_2011-11-28_1.pdf

Opel   Ampera   2012   http://www.opel-infos.de/infomaterial/75_13_2014-06-02_1.pdf   Opel Astra   http://www.opel-infos.de/infomaterial/73_13_2011-11-28_1.pdf

Toyota   Prius III 2012   https://www.adac.de/_ext/itr/tests/Autotest/AT4277_Toyota_Prius_1_8_Hybrid/Toyota_Prius_1_8_Hybrid.pdf   Toyota Auris   file://Toyota-Auris-2015-PDB_tcm-17-484491.pdf

Opel   Ampera   2013   http://www.opel-infos.de/infomaterial/75_13_2014-06-02_1.pdf   Opel Astra   http://www.opel-infos.de/infomaterial/73_13_2011-11-28_1.pdf

BMW   i3 60 Ah with range extender 2013   http://www.bmw.de/dam/brandBM/marketDE/countryDE/newvehicles/allfacts/pricelist/BMW_i3_Preisliste.pdf?download.1431068004689.pdf   BMW 320i   file://Preise_BMW_3er_Limousine_M%C3%A4rz_2015.pdf

Porsche   Panamera S E-Hybrid   2013   file://Panamera%20Preisliste%20herunterladen%20(PDF%3B%201,7%20MB).pdf   Porsche Panamera 4   file://Panamera%20Preisliste%20herunterladen%20(PDF%3B%201,7%20MB).pdf

Toyota   Prius III 2013   https://www.adac.de/_ext/itr/tests/Autotest/AT4277_Toyota_Prius_1_8_Hybrid/Toyota_Prius_1_8_Hybrid.pdf   Toyota Auris   file://Toyota-Auris-2015-PDB_tcm-17-484491.pdf

Audi   A3 e-tron 2014   https://www.audi.de/bin/dpu-de/pdf?context=nemo-de:de&ids=default-a3sbetron   Audi A3   http://www.audi-partner.de/etc/medialib/ngw/product/pdf/preislisten_my_2014.Par.0043.File.pdf/preisliste_a3-limousine.pdf

Opel   Ampera   2014   http://www.opel-infos.de/infomaterial/75_13_2014-06-02_1.pdf   Opel Astra   http://www.opel-infos.de/infomaterial/73_13_2014-12-15_1.pdf

Porsche   Cayenne S E-Hybrid   2014   http://www.porsche.com/germany/models/cayenne/cayenne-s-e-hybrid/   Porsche Cayenne S   http://www.porsche.com/germany/models/cayenne/cayenne-s/

VW	Golf GTE	2014	file://golf-gte-preisliste.pdf	VW Golf	file://golf_preisliste%20.pdf

BMW i3 60 Ah with range extender 2014 http://www.bmw.de/dam/brandBM/marketDE/countryDE/newvehicles/all-facts/pricelist/BMW_i3_Preisliste.pdf?download.1431068004689.pdf BMW 320i file://Preise_BMW_3er_Limousine_M%C3%A4rz_2015.pdf

BMW i8 2014 http://bmw-i8.club/wp-content/uploads/2015/12/201403_BMW_i8_Spezifikationen_D.pdf Audi r8 http://www.audi-hamburg.de/content/dam/ngw/product/pdf/my-2015/preislisten/preisliste_r8-coupe.pdf

Mitsubishi Outlander PHEV 2014 http://www.autohaus-siebrecht.de/sgdata/sg7338_9865.pdf Mitsubishi Outlander 2.0 MIVEC 4WD http://gebr-mueller.de/fileadmin/user_upload/PDF_Prospekte/mitsubishi_outlander_produktprospekt.pdf

Porsche Panamera E-Hybrid 2014 file://Panamera%20Preisliste%20herunterladen%20(PDF%3B%201,7%20MB).pdf Porsche Panamera 4 file://Panamera%20Preisliste%20herunterladen%20(PDF%3B%201,7%20MB).pdf

Toyota Prius III 2014 https://www.adac.de/_ext/itr/tests/Autotest/AT4277_Toyota_Prius_1_8_Hybrid/Toyota_Prius_1_8_Hybrid.pdf Toyota Auris file://Toyota-Auris-2015-PDB_tcm-17-484491.pdf

Mercedes-Benz S 500 e 2014 file://Preisliste_S-Klasse_Limousine_160812.pdf Mercedes-Benz S 500 http://www.mercedes-benz.de/content/germany/mpc/mpc_germany_website/de/home_mpc/passengercars/home/new_cars/models/s-class/w222/facts_/technicaldata/models.html#_int_passengercars:home:model-navi:models

Audi A3 e-tron 2015 https://www.audi.de/bin/dpu-de/pdf?context=nemo-de:de&ids=default-a3sbetron Audi A3 http://www.audi-partner.de/etc/medialib/ngw/product/pdf/preislisten_my_2014.Par.0043.File.pdf/preisliste_a3-limousine.pdf

Mercedes-Benz C 350 e Kombi 2015 file://S_205_C-T_PL_2016_ONLINE_211016.pdf http://www.mercedes-benz.de/content/germany/mpc/mpc_germany_website/de/home_mpc/passengercars/home/new_cars/models/c-class/s205/facts/technicaldata/models.html Mercedes-Benz C300 Kombi http://www.mercedes-benz.de/content/media_library/germany/mpc_germany/de/mercedes-benz_deutschland/pkw_emb_nextgen/neufahrzeuge/c-klasse/preislisten_pdf/pkw_c-klasse_limousine1.object-Single-MEDIA.tmp/W_205_C-Lim_PL_2016_ONLINE_211016.pdf

Mercedes-Benz C 350 e Limousine 2015 http://www.mercedes-benz.de/content/media_library/germany/mpc_germany/de/mercedes-benz_deutschland/pkw_emb_nextgen/neufahrzeuge/c-klasse/preislisten_pdf/pkw_c-klasse_limousine1.object-Single-MEDIA.tmp/W_205_C-Lim_PL_2016_ONLINE_211016.pdf Mercedes-Benz C300 Limousine http://www.mercedes-benz.de/content/media_library/germany/mpc_germany/de/mercedes-benz_deutschland/pkw_emb_nextgen/neufahrzeuge/c-klasse/preislisten_pdf/pkw_c-klasse_limousine1.object-Single-MEDIA.tmp/W_205_C-Lim_PL_2016_ONLINE_211016.pdf

Porsche Cayenne S E-Hybrid 2015 http://www.porsche.com/germany/models/cayenne/cayenne-s-e-hybrid/ Porsche Cayenne S http://www.porsche.com/germany/models/cayenne/cayenne-s/

Mercedes-Benz GLC 350e 2015 file://Preisliste_GLC_SUV_161006.pdf Mercedes-Benz GLC 350 d 4MATIC file://Preisliste_GLC_SUV_161006.pdf

Mercedes-Benz GLE 500 e 4MATIC 2015 file://Preisliste_GLE_166_161006%20(1).pdf Mercedes-Benz GLE 500 4MATIC file://Preisliste_GLE_166_161006%20.pdf

VW	Golf GTE		2015	file://golf-gte-preisliste.pdf	VW Golf	file://golf_preisliste%20.pdf

BMW i3 60 Ah with rage extender 2015 http://www.bmw.de/dam/brandBM/marketDE/countryDE/newvehicles/all-facts/pricelist/BMW_i3_Preisliste.pdf?download.1431068004689.pdf BMW 320i file://Preise_BMW_3er_Limousine_M%C3%A4rz_2015.pdf

BMW i8 2015 http://bmw-i8.club/wp-content/uploads/2015/12/201403_BMW_i8_Spezifikationen_D.pdf Audi r8 http://www.audi-hamburg.de/content/dam/ngw/product/pdf/my-2015/preislisten/preisliste_r8-coupe.pdf

Mitsubishi Outlander PHEV 2015 http://www.autohaus-siebrecht.de/sgdata/sg7338_9865.pdf Mitsubishi Outlander 2.0 MIVEC 4WD http://gebr-mueller.de/fileadmin/user_upload/PDF_Prospekte/mitsubishi_outlander_produktprospekt.pdf

Porsche Panamera E-Hybrid 2015 file://Panamera%20Preisliste%20herunterladen%20(PDF%3B%201,7%20MB).pdf Porsche Panamera 4 file://Panamera%20Preisliste%20herunterladen%20(PDF%3B%201,7%20MB).pdf

VW Passat GTE 2015 file://passat-gte-passat-gte-variant_preisliste.pdf VW Passat TDI Bluemotion file://passat-passatvariant_preisliste.pdf

VW Passat Variant GTE 2015 file://passat-gte-passat-gte-variant_preisliste.pdf VW Passat Variant TDI Bluemotion file://passat-passatvariant_preisliste.pdf

Toyota Prius III 2015 https://www.adac.de/_ext/itr/tests/Autotest/AT4277_Toyota_Prius_1_8_Hybrid/Toyota_Prius_1_8_Hybrid.pdf Toyota Auris file://Toyota-Auris-2015-PDB_tcm-17-484491.pdf

Mercedes-Benz S 500 e 2015 http://www.mercedes-benz.de/content/germany/mpc/mpc_germany_website/de/home_mpc/passengercars/home/new_cars/models/s-class/w222/facts_/technicaldata/models.html#_int_passengercars:home:model-navi:models Mercedes-Benz S 500 http://www.mercedes-benz.de/content/germany/mpc/mpc_germany_website/de/home_mpc/passengercars/home/new_cars/models/s-class/w222/facts_/technicaldata/models.html#_int_passengercars:home:model-navi:models

BMW X5 xDrive40e 2015 file://der_bmw_x5_xdrive40e.pdf BMW X5 xDrive 35i file://BMW_X5_Preisliste.pdf

Volvo XC90 T8 Twin Engine 2015 http://www.volvocars.com/de/modelle/neuwagen/xc90/t8-twin-engine Volvo XC90 T6 AWD Momentum file:///Preisliste%20Volvo%20XC90%20MY17.pdf

BMW 225xe Active Tourer 2016 file:///BMW_2er_Coupe_Preisliste%20.pdf BMW 225d file://BMW_2er_Coupe_Preisliste%20.pdf

BMW 330e 2016 http://www.motorsport-total.com/auto/news/2016/08/bmw-330e-spagat-zwischen-umwelt-und-leistung-16081901.html BMW 330i file://BMW_3er_Limousine_Preisliste%20.pdf

BMW 740e 2016 file:///BMW_7er_Preisliste%20.pdf https://www.press.bmwgroup.com/deutschland/article/detail/T0261925DE/der-neue-bmw-740e-iperformance-der-neue-bmw-740le-iperformance-der-neue-bmw-740le-xdrive-iperformance?language=de BMW 740i file://BMW_7er_Preisliste%20.pdf

Audi A3 e-tron 2016 https://www.audi.de/bin/dpu-de/pdf?context=nemo-de:de&ids=default-a3sbetron Audi A3 http://www.audi-partner.de/etc/medialib/ngw/product/pdf/preislisten_my_2014.Par.0043.File.pdf/preisliste_a3-limousine.pdf

Mercedes-Benz C 350 e Kombi 2016 file://S_205_C-T_PL_2016_ONLINE_211016.pdf http://www.mercedes-benz.de/content/germany/mpc/mpc_germany_website/de/home_mpc/passengercars/home/new_cars/models/c-class/s205/facts/technicaldata/models.html Mercedes-Benz C300 Kombi http://www.mercedes-benz.de/content/media_library/germany/mpc_germany/de/mercedes-benz_deutschland/pkw_emb_nextgen/neufahrzeuge/c-klasse/preislisten_pdf/pkw_c-klasse_limousine1.object-Single-MEDIA.tmp/W_205_C-Lim_PL_2016_ONLINE_211016.pdf

Mercedes-Benz C 350 e Limousine 2016 http://www.mercedes-benz.de/content/media_library/germany/mpc_germany/de/mercedes-benz_deutschland/pkw_emb_nextgen/neufahrzeuge/c-klasse/preislisten_pdf/pkw_c-klasse_limousine1.object-Single-MEDIA.tmp/W_205_C-Lim_PL_2016_ONLINE_211016.pdf Mercedes-Benz C300 Limousine http://www.mercedes-benz.de/content/media_library/germany/mpc_germany/de/mercedes-benz_deutschland/pkw_emb_nextgen/neufahrzeuge/c-klasse/preislisten_pdf/pkw_c-klasse_limousine1.object-Single-MEDIA.tmp/W_205_C-Lim_PL_2016_ONLINE_211016.pdf

Porsche Cayenne S E-Hybrid 2016 http://www.porsche.com/germany/models/cayenne/cayenne-s-e-hybrid/ Porsche Cayenne S http://www.porsche.com/germany/models/cayenne/cayenne-s/

Mercedes-Benz GLC 350e 2016 http://www.mercedes-benz.de/content/germany/mpc/mpc_germany_website/de/home_mpc/passengercars/home/new_cars/models/glc/x253/facts/technicaldata/models.html file://glc_de_06-2016.pdf Mercedes-Benz GLC 350 d 4MATIC file://Preisliste_GLC_SUV_161006.pdf

Mercedes-Benz GLE 500 e 4MATIC 2016 http://www.mercedes-benz.de/content/germany/mpc/mpc_germany_website/de/home_mpc/passengercars/home/new_cars/models/gle-class/gle_suv/facts/technicaldata/models.html https://www.daimler.com/bilder/nachhaltigkeit/produkt/neu-umweltzertifikate/daimler-umweltzertifikat-compact-mb-gle.pdf Mercedes-Benz GLE 500 4MATIC http://www.mercedes-benz.de/content/germany/mpc/mpc_germany_website/de/home_mpc/passengercars/home/new_cars/models/gle-class/gle_coupe/facts/technicaldata/models.html

VW Golf GTE 2016 file://golf-gte-preisliste.pdf VW Golf file://golf_preisliste%20.pdf

BMW i3 60 Ah with Rangeextender 2016 http://www.bmw.de/dam/brandBM/marketDE/countryDE/newvehicles/allfacts/pricelist/BMW_i3_Preisliste.pdf?download.1431068004689.pdf BMW 320i file://Preise_BMW_3er_Limousine_M%C3%A4rz_2015.pdf

BMW i3 94 Ah with Rangeextender 2016 http://www.bmw.de/dam/brandBM/marketDE/countryDE/newvehicles/allfacts/pricelist/BMW_i3_Preisliste.pdf?download.1431068004689.pdf BMW 320i file://Preise_BMW_3er_Limousine_M%C3%A4rz_2015.pdf

BMW i8 2016 http://bmw-i8.club/wp-content/uploads/2015/12/201403_BMW_i8_Spezifikationen_D.pdf Audi r8 http://www.audi-hamburg.de/content/dam/ngw/product/pdf/my-2015/preislisten/preisliste_r8-coupe.pdf

Kia Optima Plug-in-Hybrid 2016 http://www.kia.com/de/modelle/optima-phev/ Kia Optima http://www.kia.com/de/modelle/optima/

Mitsubishi Outlander PHEV 2016 http://www.autohaus-siebrecht.de/sgdata/sg7338_9865.pdf Mitsubishi Outlander 2.0 MIVEC 4WD http://gebr-mueller.de/fileadmin/user_upload/PDF_Prospekte/mitsubishi_outlander_produktprospekt.pdf

Porsche	Panamera E-Hybrid	2016	file://Panamera%20Preisliste%20herunterladen%20(PDF%3B%201,7%20MB).pdf	
			Porsche Panamera 4	file:///Panamera%20Preisliste%20herunterladen%20(PDF%3B%201,7%20MB).pdf
VW	Passat GTE	2016	file://passat-gte-passat-gte-variant_preisliste.pdf	VW Passat TDI Bluemotion file://passat-passatvariant_preisliste.pdf
VW	Passat Variant GTE	2016	file://passat-gte-passat-gte-variant_preisliste.pdf	VW Passat Variant TDI Bluemotion file://passat-passatvariant_preisliste.pdf
Toyota	Prius IV 2016	file://Toyota-Prius-PDB-2016-02_tcm-17-100397.pdf	Toyota Auris	file:///Toyota-Auris-2015-PDB_tcm-17-484491.pdf
Audi	Q7 e-tron 2016	https://www.audi-mediacenter.com/de/audi-q7-e-tron-quattro-tdi-66	Audi Q7 https://www.audi.de/dam/nemo/models/misc/pdf/my-2015/preislisten/preisliste_q7.pdf	
Mercedes-Benz	S 500 e 2016	file://Preisliste_S-Klasse_Limousine_160812.pdf	Mercedes-Benz S 500 http://www.mercedes-benz.de/content/germany/mpc/mpc_germany_website/de/home_mpc/passengercars/home/new_cars/models/s-class/w222/facts_/technicaldata/models.html#_int_passengercars:home:model-navi:models	
BMW	X5 xDrive40e	2016	file://der_bmw_x5_xdrive40e.pdf	BMW X5 xDrive 35i file://BMW_X5_Preisliste.pdf
Volvo	XC90 T8 Twin Engine	2016	http://www.volvocars.com/de/modelle/neuwagen/xc90/t8-twin-engine	
	Volvo XC90 T6 AWD Momentum		file://Preisliste%20Volvo%20XC90%20MY17.pdf	

List 5: yearly maintenance and fix costs of BEVs, PHEVs and CVs, all obtained from ADAC (2017), ADAC Autokosten-Rechner, www.adac.de/autokosten. Retrieved until 31. January 2017

BEVS, yearly costs

https://www.adac.de/infotestrat/autodatenbank/autokosten/detail.aspx?KFZID=237870&activeTab=3&info=Nissan+Leaf+(24+kWh)+Visia+(inkl.+Batterie)

https://www.adac.de/infotestrat/autodatenbank/autokosten/detail.aspx?KFZID=237871&activeTab=3&info=Nissan+Leaf+(24+kWh)+Acenta+(inkl.+Batterie)

https://www.adac.de/infotestrat/autodatenbank/autokosten/detail.aspx?KFZID=237872&activeTab=3&info=Nissan+Leaf+(24+kWh)+Tekna+(inkl.+Batterie)

https://www.adac.de/infotestrat/autodatenbank/autokosten/detail.aspx?KFZID=238074&activeTab=3&info=Ford+Focus+Electric+

https://www.adac.de/infotestrat/autodatenbank/autokosten/detail.aspx?KFZID=238113&activeTab=3&info=BMW+i3+(60+Ah)+

https://www.adac.de/infotestrat/autodatenbank/autokosten/detail.aspx?KFZID=240967&activeTab=3&info=Mitsubishi+Electric+Vehicle+

https://www.adac.de/infotestrat/autodatenbank/autokosten/detail.aspx?KFZID=249417&activeTab=3&info=Mercedes+B+250+e+

https://www.adac.de/infotestrat/autodatenbank/autokosten/detail.aspx?KFZID=250860&activeTab=3&info=Nissan+Leaf+(30+kWh)+Acenta+(inkl.+Batterie)

https://www.adac.de/infotestrat/autodatenbank/autokosten/detail.aspx?KFZID=250862&activeTab=3&info=Nissan+Leaf+(30+kWh)+Tekna+(inkl.+Batterie)
https://www.adac.de/infotestrat/autodatenbank/autokosten/detail.aspx?KFZID=254800&activeTab=3&info=Nissan+e-NV200+Evalia+Tekna+(5-Sitzer)+(inkl.+Batterie)
https://www.adac.de/infotestrat/autodatenbank/autokosten/detail.aspx?KFZID=254864&activeTab=3&info=Tesla+Model+S+90D+
https://www.adac.de/infotestrat/autodatenbank/autokosten/detail.aspx?KFZID=254883&activeTab=3&info=Tesla+Model+X+75D+
https://www.adac.de/infotestrat/autodatenbank/autokosten/detail.aspx?KFZID=254884&activeTab=3&info=Tesla+Model+X+90D+
https://www.adac.de/infotestrat/autodatenbank/autokosten/detail.aspx?KFZID=255581&activeTab=3&info=BMW+i3+(94+Ah)+
https://www.adac.de/infotestrat/autodatenbank/autokosten/detail.aspx?KFZID=257041&activeTab=3&info=Tesla+Model+S+60+
https://www.adac.de/infotestrat/autodatenbank/autokosten/detail.aspx?KFZID=259140&activeTab=3&info=KIA+Soul+EV+Play
https://www.adac.de/infotestrat/autodatenbank/autokosten/detail.aspx?KFZID=259955&activeTab=3&info=Hyundai+IONIQ+Elektro+Trend
https://www.adac.de/infotestrat/autodatenbank/autokosten/detail.aspx?KFZID=261164&activeTab=3&info=VW+e-up!+
https://www.adac.de/infotestrat/autodatenbank/autokosten/detail.aspx?KFZID=261846&activeTab=3&info=Citroen+C-Zero+
https://www.adac.de/infotestrat/autodatenbank/autokosten/detail.aspx?KFZID=261847&activeTab=3&info=Peugeot+iOn+Active
https://www.adac.de/infotestrat/autodatenbank/autokosten/detail.aspx?KFZID=266739&activeTab=3&info=Renault+Zoe+(41+kWh)+Life+(inkl.+Batterie)
https://www.adac.de/infotestrat/autodatenbank/autokosten/detail.aspx?KFZID=268176&activeTab=3&info=Tesla+Model+X+100D+

BEV - CV Equivalents, yearly costs:
https://www.adac.de/infotestrat/autodatenbank/autokosten/detail.aspx?KFZID=242614&activeTab=3&info=Ford+Focus+1.0+EcoBoost+Start%2fStopp+Trend
https://www.adac.de/infotestrat/autodatenbank/autokosten/detail.aspx?KFZID=242858&activeTab=3&info=Opel+Corsa+1.2+Selection
https://www.adac.de/infotestrat/autodatenbank/autokosten/detail.aspx?KFZID=243338&activeTab=3&info=Mercedes+C+63+AMG+AMG+SPEEDSHIFT+MCT
https://www.adac.de/infotestrat/autodatenbank/autokosten/detail.aspx?KFZID=244178&activeTab=3&info=smart+fortwo+coup%C3%A9+1.0+
https://www.adac.de/infotestrat/autodatenbank/autokosten/detail.aspx?KFZID=245012&activeTab=3&info=Opel+KARL+1.0+Selection
https://www.adac.de/infotestrat/autodatenbank/autokosten/detail.aspx?KFZID=245901&activeTab=3&info=Peugeot+Partner+Tepee+VTi+98+Access
https://www.adac.de/infotestrat/autodatenbank/autokosten/detail.aspx?KFZID=246999&activeTab=3&info=BMW+320i+
https://www.adac.de/infotestrat/autodatenbank/autokosten/detail.aspx?KFZID=247036&activeTab=3&info=BMW+340i+Advantage

https://www.adac.de/infotestrat/autodatenbank/autokosten/detail.aspx?KFZID=248214&activeTab=3&info=Citroen+C1+VTi+68+Start
https://www.adac.de/infotestrat/autodatenbank/autokosten/detail.aspx?KFZID=249134&activeTab=3&info=Fiat+500+1.2+8V+Pop
https://www.adac.de/infotestrat/autodatenbank/autokosten/detail.aspx?KFZID=249668&activeTab=3&info=Nissan+Evalia+16V+110+Tekna
https://www.adac.de/infotestrat/autodatenbank/autokosten/detail.aspx?KFZID=250321&activeTab=3&info=Toyota+Auris+1.33+
https://www.adac.de/infotestrat/autodatenbank/autokosten/detail.aspx?KFZID=256051&activeTab=3&info=Audi+SQ7+TDI+quattro+tiptronic
https://www.adac.de/infotestrat/autodatenbank/autokosten/detail.aspx?KFZID=256051&activeTab=3&info=Audi+SQ7+TDI+quattro+tiptronic
https://www.adac.de/infotestrat/autodatenbank/autokosten/detail.aspx?KFZID=256051&activeTab=3&info=Audi+SQ7+TDI+quattro+tiptronic
https://www.adac.de/infotestrat/autodatenbank/autokosten/detail.aspx?KFZID=258058&activeTab=3&info=Renault+Clio+ENERGY+TCe+90+Intens
https://www.adac.de/infotestrat/autodatenbank/autokosten/detail.aspx?KFZID=266207&activeTab=3&info=VW+Golf+1.6+TDI+BMT+Trendline+(ab+03%2f17)

PHEV, yearly costs

https://www.adac.de/infotestrat/autodatenbank/autokosten/detail.aspx?KFZID=238114&activeTab=3&info=BMW+i3+(60+Ah)+(inkl.+Range+Extender)+
https://www.adac.de/infotestrat/autodatenbank/autokosten/detail.aspx?KFZID=238201&activeTab=3&info=BMW+i8+
https://www.adac.de/infotestrat/autodatenbank/autokosten/detail.aspx?KFZID=242363&activeTab=3&info=Porsche+Cayenne+S+E-Hybrid+Tiptronic+S
https://www.adac.de/infotestrat/autodatenbank/autokosten/detail.aspx?KFZID=243353&activeTab=3&info=Volvo+XC90+T8+Twin+Engine+Momentum+AWD+Geartronic
https://www.adac.de/infotestrat/autodatenbank/autokosten/detail.aspx?KFZID=245149&activeTab=3&info=Mercedes+C+350+e+T-Modell+Avantgarde+7G-TRONIC+PLUS
https://www.adac.de/infotestrat/autodatenbank/autokosten/detail.aspx?KFZID=245168&activeTab=3&info=Mercedes+C+350+e+Avantgarde+7G-TRONIC+PLUS
https://www.adac.de/infotestrat/autodatenbank/autokosten/detail.aspx?KFZID=245935&activeTab=3&info=Mercedes+GLE+500+e+4MATIC+7G-TRONIC+PLUS
https://www.adac.de/infotestrat/autodatenbank/autokosten/detail.aspx?KFZID=246092&activeTab=3&info=Mercedes+S+500+e+lang+7G-TRONIC+PLUS
https://www.adac.de/infotestrat/autodatenbank/autokosten/detail.aspx?KFZID=248000&activeTab=3&info=BMW+X5+xDrive40e+iPerformance+Steptronic
https://www.adac.de/infotestrat/autodatenbank/autokosten/detail.aspx?KFZID=248181&activeTab=3&info=VW+Passat+GTE+DSG
https://www.adac.de/infotestrat/autodatenbank/autokosten/detail.aspx?KFZID=248297&activeTab=3&info=BMW+740e+iPerformance+Steptronic
https://www.adac.de/infotestrat/autodatenbank/autokosten/detail.aspx?KFZID=250565&activeTab=3&info=BMW+225xe+Active+Tourer+Advantage+xDrive+Steptronic
https://www.adac.de/infotestrat/autodatenbank/autokosten/detail.aspx?KFZID=250589&activeTab=3&info=BMW+330e+iPerformance+Advantage+Steptronic

https://www.adac.de/infotestrat/autodatenbank/autokosten/detail.aspx?KFZID=250994&activeTab=3&info=Mitsubishi+Outlander+Plug-In+Hybrid+Basis+4WD
https://www.adac.de/infotestrat/autodatenbank/autokosten/detail.aspx?KFZID=251369&activeTab=3&info=Audi+Q7+e-tron+quattro+tiptronic
https://www.adac.de/infotestrat/autodatenbank/autokosten/detail.aspx?KFZID=255014&activeTab=3&info=Mercedes+GLC+350+e+4MATIC+7G-TRONIC+PLUS
https://www.adac.de/infotestrat/autodatenbank/autokosten/detail.aspx?KFZID=255014&activeTab=3&info=Mercedes+GLC+350+e+4MATIC+7G-TRONIC+PLUS
https://www.adac.de/infotestrat/autodatenbank/autokosten/detail.aspx?KFZID=255582&activeTab=3&info=BMW+i3+(94+Ah)+(inkl.+Range+Extender)+
https://www.adac.de/infotestrat/autodatenbank/autokosten/detail.aspx?KFZID=258227&activeTab=3&info=Audi+A3+Sportback+e-tron+S+tronic
https://www.adac.de/infotestrat/autodatenbank/autokosten/detail.aspx?KFZID=261129&activeTab=3&info=KIA+Optima+2.0+GDI+Plugin-Hybrid+Attract
https://www.adac.de/infotestrat/autodatenbank/autokosten/detail.aspx?KFZID=261790&activeTab=3&info=Porsche+Panamera+4+E-Hybrid+PDK+(ab+04%2f17)
https://www.adac.de/infotestrat/autodatenbank/autokosten/detail.aspx?KFZID=266948&activeTab=3&info=VW+Golf+GTE+DSG+(ab+03%2f17)

PHEV - CV Equivalents, yearly costs
https://www.adac.de/infotestrat/autodatenbank/autokosten/detail.aspx?KFZID=237123&activeTab=3&info=BMW+X5+xDrive35i+Steptronic
https://www.adac.de/infotestrat/autodatenbank/autokosten/detail.aspx?KFZID=241904&activeTab=3&info=VW+Passat+2.0+TDI+BMT+Trendline
https://www.adac.de/infotestrat/autodatenbank/autokosten/detail.aspx?KFZID=241912&activeTab=3&info=VW+Passat+Variant+1.4+TSI+BMT+ACT+Comfortline
https://www.adac.de/infotestrat/autodatenbank/autokosten/detail.aspx?KFZID=242362&activeTab=3&info=Porsche+Cayenne+S+Tiptronic+S
https://www.adac.de/infotestrat/autodatenbank/autokosten/detail.aspx?KFZID=243350&activeTab=3&info=Volvo+XC90+T6+Momentum+AWD+Geartronic
https://www.adac.de/infotestrat/autodatenbank/autokosten/detail.aspx?KFZID=245651&activeTab=3&info=Audi+Q7+3.0+TFSI+quattro+tiptronic
https://www.adac.de/infotestrat/autodatenbank/autokosten/detail.aspx?KFZID=246999&activeTab=3&info=BMW+320i+
https://www.adac.de/infotestrat/autodatenbank/autokosten/detail.aspx?KFZID=247021&activeTab=3&info=BMW+330i+Advantage
https://www.adac.de/infotestrat/autodatenbank/autokosten/detail.aspx?KFZID=247586&activeTab=3&info=Audi+R8+Coup%C3%A9+5.2+FSI+V10+quattro+S+tronic
https://www.adac.de/infotestrat/autodatenbank/autokosten/detail.aspx?KFZID=248291&activeTab=3&info=BMW+740i+Steptronic
https://www.adac.de/infotestrat/autodatenbank/autokosten/detail.aspx?KFZID=250321&activeTab=3&info=Toyota+Auris+1.33+
https://www.adac.de/infotestrat/autodatenbank/autokosten/detail.aspx?KFZID=250986&activeTab=3&info=Mercedes+GLE+500+4MATIC+9G-TRONIC
https://www.adac.de/infotestrat/autodatenbank/autokosten/detail.aspx?KFZID=250999&activeTab=3&info=Mitsubishi+Outlander+2.0+ClearTec+Plus+4WD+CVT-Automatik

https://www.adac.de/infotestrat/autodatenbank/autokosten/detail.aspx?KFZID=251012&activeTab=3&info=KIA+Optima+2.0+Edition+7

https://www.adac.de/infotestrat/autodatenbank/autokosten/detail.aspx?KFZID=252611&activeTab=3&info=Mercedes+S+500+9G-TRONIC

https://www.adac.de/infotestrat/autodatenbank/autokosten/detail.aspx?KFZID=262027&activeTab=3&info=Mercedes+C+300+Avantgarde+9G-TRONIC

https://www.adac.de/infotestrat/autodatenbank/autokosten/detail.aspx?KFZID=262055&activeTab=3&info=Mercedes+C+300+T-Modell+Avantgarde+9G-TRONIC

https://www.adac.de/infotestrat/autodatenbank/autokosten/detail.aspx?KFZID=262085&activeTab=3&info=Mercedes+GLC+350+d+4MATIC+9G-TRONIC

https://www.adac.de/infotestrat/autodatenbank/autokosten/detail.aspx?KFZID=263677&activeTab=3&info=Porsche+Panamera+4+PDK

https://www.adac.de/infotestrat/autodatenbank/autokosten/detail.aspx?KFZID=266234&activeTab=3&info=VW+Golf+1.4+TSI+BMT+Highline+(ab+03%2f17)

Appendix A: Overview of collected data

Table 7: Average fuel and electricity prices (SB, 2017)

Gasoline		Diesel		Electricity	
Year	Price [€/l]	Year	Price [€/l]	Year	Price [€/kWh]
2010	141.5	2010	122.4	2010	23.69
2011	155.4	2011	141.9	2011	25.23
2012	164.6	2012	148.9	2012	25.89
2013	159.2	2013	142.8	2013	28.84
2014	152.8	2014	135	2014	29.14
2015	139.4	2015	117.1	2015	28.7
2016	129.6	2016	107.21	2016	28.8
Avg	148.93		130.76		27.18

Table 8: Yearly inflation rates for motor cars in Germany (Eurostat, 2017)

Year	HICP Motor Cars (2015 = 100) - annual data (average index and rate of change)
1996	86.70
1997	86.50
1998	87.70
1999	88.40
2000	88.60
2001	89.50
2002	90.80
2003	91.60
2004	92.70

Year	HICP Motor Cars (2015 = 100) - annual data (average index and rate of change)
2005	93.20
2006	94.60
2007	97.60
2008	98.40
2009	98.30
2010	98.20
2011	98.60
2012	98.80
2013	98.70
2014	99.00
2015	100.00
2016	100.25
2017	100.50

Year	Divergenz on average [%]	Coefficient
2010	24.8	1.248
2011	28.4	1.284
2012	30.5	1.305
2013	36.0	1.36
2014	40.0	1.4
2015	42.0	1.42
2016	42.0	1.42

Table 9: Divergence between certified energy consumption and real world energy consumption of CVs (ICCT, 2016b)

Table 10: Divergence between certified energy consumption and real world energy consumption of PHEVs (Tietge et al., 2015)

Divergence on Average [%]	Coefficient	n	Source
265	2.65	130	ICCT - HONESTJOHN.CO.UK (UNITED KINGDOM)
211	2.11	995	ICCT - CLEANER CAR CONTRACTS
290	2.9	3	ICCT
250	2.5	7	ICCT
217.63	2.18	1135	

Table 11: Production and cumulated production of electric vehicles worldwide until 2016 (ZWS, 2016; Weiss et al., 2012)

Year	E-Cars worldwide cumulated (Source: ZWS, M. Weiss)	New registrations worldwide (Source: ZWS)	E-Cars worldwide cumulated (Source: ZWS)	E-Cars worldwide cumulated (Source: ZWS) (1000 Units)	E-Cars worldwide cumulated (Source: M. Weiss)	E-Cars worldwide cumulated until 2008. difference M.Weiss to ZWS (Source: M. Weiss)
2008	52,410					52,410
2009	54,390	1,980	1,980	1.98		
2010	59,260	4,870	6,850	6.85		
2011	67,900	8,640	15,490	15.49	67,900	
2012	122,440	54,540	70,030	70.03		
2013	252,420	129,980	200,010	200.01		
2014	471,760	219,340	419,350	419.35		
2015	800,790	329,030	748,380	748.38		
2016	1,352,220	551,430	1,299,810	1,299.81		

Table 12: Price data of BEVs and their equivalent CVs

Manufacturer	Model	Year	Power [KW]	Battery [kWh]	Net Price [€]	Real Price Net [€$_{2015}$]	Real Specific Price 1, Net [€$_{2015}$/kWh]	Real Specific price 2, Net [€$_{2015}$/kW]	Equivalent ICE Car	Power [KW]	Net Price [€]	Real Price with tax [€$_{2015}$]	Real Price Net [€$_{2015}$]	Real Specific price 2, Net [€$_{2015}$/kW]
Mitsubishi	i-MiEV	2010	49	16	29403	29942	1871	611.07	Citroën C1	50	7134	8646	7265	145
Fiat	Karabag 500 E	2010	30	11	51158	52096	4736	1736.52	Fiat 500 Opel Agila B / Suzuki Splash 2010	51	10294	12475	10483	206
German E-Cars	Stromos	2010	56	19.2	35286	35932	1871	641.65	Citroën C1	48	8639	10468	8797	183
Tazzari	Zero	2010	15	12.3	20160	20529	1669	1368.61	Citroën C1	50	7134	8646	7265	145
Mitsubishi	i-MiEV	2011	49	16	29403	29821	1864	608.59	Citroën C1	50	7134	8611	7236	145
Fiat	Karabag 500 E	2011	28	11	36134	36648	3332	1308.84	Fiat 500	51	10412	12566	10560	207
Aixam	Mega e-City	2011	4	13	23517	23851	1835	5962.68	Aixam / Ligier	5	8403	10142	8523	1705
German E-Cars	Stromos	2011	56	19.2	35286	35787	1864	639.05	Opel Agila B / Suzuki Splash	50	8908	10751	9034	181
Tazzari	Zero	2011	15	12.3	20160	20446	1662	1363.06	Citroën C1	50	7134	8611	7236	145
German E-Cars	CETOS	2012	60	19.2	36750	37196	1937	619.94	Opel Corsa	51	9937	11969	10058	197
Citroën	C-Zero	2012	49	14.5	24700	25000	1724	510.20	Citroën C1	50	7050	8492	7136	143
Smart	ED	2012	55	17.6	19899	20141	1144	366.20	Smart fortwo	52	9105	10967	9216	177
Peugeot	iOn	2012	49	14.5	24700	25000	1724	510.20	Citroën C1	50	7050	8492	7136	143
Mitsubishi	i-MiEV	2012	49	16	24622	24921	1558	508.59	Citroën C1	50	7050	8492	7136	143
Fiat	Karabag 500 E	2012	20	11	36134	36573	3325	1828.67	Fiat 500	51	9748	11741	9866	193
Nissan	Leaf Visia	2012	80	24	31084	31462	1311	393.27	Ford Focus 1.0 EcoBoost	92	16975	20445	17181	187
German E-Cars	Stromos	2012	56	19.2	26471	26792	1395	478.43	Opel Agila B / Suzuki Splash	50	9063	10916	9173	183
Tazzari	Zero	2012	15	12.3	20160	20405	1659	1360.30	Citroën C1	50	7050	8492	7136	143

Manufacturer	Model	Year	Power [KW]	Battery [kWh]	Net Price [€]	Real Price Net [€$_{2015}$]	Real Specific Price 1, Net [€$_{2015}$/kWh]	Real Specific price 2, Net [€$_{2015}$/kW]	Equivalent ICE Car	Power [KW]	Net Price [€]	Real Price with tax [€$_{2015}$]	Real Price Net [€$_{2015}$]	Real Specific price 2, Net [€$_{2015}$/kW]
German E-Cars	CETOS	2013	60	19.2	36750	37234	1939	620.57	Opel Corsa	51	9992	12047	10123	198
Citroën	C-Zero	2013	49	14.5	24700	25025	1726	510.72	Citroën C1	50	7286	8784	7382	148
Smart	ED	2013	55	17.6	19899	20161	1146	366.57	Smart fortwo	52	9155	11039	9276	178
VW	e-up	2013	60	18.7	22605	22903	1225	381.71	VW up! (cheer up)	55	10441	12589	10579	192
Ford	Focus Electric	2013	107	23	33605	34048	1480	318.20	Ford Focus 1.0 EcoBoost	92	17160	20689	17386	189
Peugeot	iOn	2013	49	14.5	24700	25025	1726	510.72	Citroën C1	50	7286	8784	7382	148
BMW	i3 60 Ah	2013	125	22	29370	29757	1353	238.05	BMW 320i	135	28067	33840	28437	211
Mitsubishi	i-MiEV	2013	49	16	24622	24946	1559	509.11	Citroën C1	50	7286	8784	7382	148
Fiat	Karabag 500 E	2013	20	11	36134	36610	3328	1830.52	Fiat 500 Ford Focus	51	9832	11854	9961	195
Nissan	Leaf Visia	2013	80	24	24950	25278	1053	315.98	1.0 EcoBoost	92	17160	20689	17386	189
Tesla	Model S	2013	225	85	68697	69602	819	309.34	BMW 335i	225	36933	44529	37419	166
Tesla	Model S 60	2013	225	60	60000	60790	1013	270.18	BMW 335i	225	36933	44529	37419	166
German E-Cars	Stromos	2013	56	19.2	26471	26819	1397	478.91	Opel Agila B / Suzuki Splash	50	9063	10927	9182	184
Tazzari	Zero	2013	15	12.3	20160	20425	1661	1361.68	Citroën C1	50	7286	8784	7382	148
Mercedes-Benz	B-Klasse Electric Drive	2014	132	28	32900	33232	1187	251.76	Mercedes-Benz B 250	155	29300	35219	29596	191
German E-Cars	CETOS	2014	60	19.2	36750	37121	1933	618.68	Opel Corsa	51	10067	12101	10169	199
Citroën	C-Zero	2014	49	14.5	24700	24949	1721	509.17	Citroën C1	51	7471	8980	7546	148
Smart	ED	2014	55	17.6	19899	20100	1142	365.46	Smart fortwo	52	9155	11005	9248	178
Volkswagen	E-Golf	2014	85	24.2	29328	29624	1224	348.52	VW Golf TSI BlueMotion Comfortline	85	19370	23283	19565	230
VW	e-up (cheer up)	2014	60	18.7	22605	22833	1221	380.56	VW up!	55	10441	12551	10547	192
Ford	Focus Electric	2014	107	23	33605	33944	1476	317.24	Ford Focus 1.0 EcoBoost	92	17277	20768	17452	190

Manufacturer	Model	Year	Power [KW]	Battery [kWh]	Net Price [€]	Real Price Net [€2015]	Real Specific Price 1, Net [€2015/kWh]	Real Specific price 2, Net [€2015/kW]	Equivalent ICE Car	Power [KW]	Net Price [€]	Real Price with tax [€2015]	Real Price Net [€2015]	Real Specific price 2, Net [€2015/kW]
Peugeot	iOn	2014	49	14.5	24700	24949	1721	509.17	Citroën C1	51	7471	8980	7546	148
BMW	i3 60 Ah	2014	125	22	29370	29666	1348	237.33	BMW 320i	135	28277	33990	28563	212
Mitsubishi	i-MiEV	2014	49	16	19992	20194	1262	412.11	Citroën C1	51	7471	8980	7546	148
Fiat	Karabag 500 E	2014	30	11	22479	22706	2064	756.87	Fiat 500 Ford Focus 1.0 EcoBoost	51	10042	12071	10143	199
Nissan	Leaf Visia	2014	80	24	24950	25202	1050	315.02		92	17277	20768	17452	190
Tesla	Model S 60	2014	225	60	54874	55428	924	246.35	BMW 335i	225	37143	44646	37518	167
Peugeot	Partner Electric L1	2014	49	22.5	26100	26364	1172	538.03	Peugeot Partner Kastenwagen	72	15101	18152	15253	212
Kia	Soul EV	2014	81	27	25874	26135	968	322.66	Kia Soul	97	14277	17162	14422	149
German E-Cars	Stromos	2014	56	19.2	26471	26738	1393	477.46	Opel Agila B	50	9063	10894	9155	183
Tazzari	Zero	2014	25	15	20471	20677	1378	827.09	Citroën C1	51	7471	8980	7546	148
Mercedes-Benz	B-Klasse Electric Drive	2015	132	28	32900	32900	1175	249.24	Mercedes-Benz B 250	155	29300	34867	29300	189
CCS Elektromobile	City-Stromerle Coupe	2015	10	12	10076	10076	840	1007.56	Aixam Miniauto L7e	15	10908	12980	10908	727
Citroën	C-Zero	2015	49	14.5	24700	24700	1703	504.08	Citroën C1	51	7471	8890	7471	146
Smart	ED	2015	55	17.6	20664	20664	1174	375.71	Smart fortwo VW Golf TSI BlueMotion Comfortline VW up! (cheer up)	52	9155	10895	9155	176
Volkswagen	E-Golf	2015	85	24.2	29328	29328	1212	345.03		85	19370	23050	19370	228
VW	e-up	2015	60	18.7	22605	22605	1209	376.75		55	10441	12425	10441	190
Ford	Focus Electric	2015	107	23	29328	29328	1275	274.09	Ford Focus 1.0 EcoBoost	92	17277	20560	17277	188
Peugeot	iOn	2015	49	14.5	15000	15000	1034	306.12	Citroën C1	51	7471	8890	7471	146
BMW	i3 60 Ah	2015	125	22	29370	29370	1335	234.96	BMW 320i	135	28782	34250	28782	213
Mitsubishi	i-MiEV	2015	49	16	19992	19992	1249	407.99	Citroën C1	51	7471	8890	7471	146
Nissan	LEAF ACENTA	2015	80	30	28538	28538	951	356.72	Ford Focus 1.0 EcoBoost	92	17277	20560	17277	188

Manufacturer	Model	Year	Power [KW]	Battery [kWh]	Net Price [€]	Real Price Net [€2015]	Real Specific Price 1, Net [€2015/kWh]	Real Specific price 2, Net [€2015/kW]	Equivalent ICE Car	Power [KW]	Net Price [€]	Real Price with tax [€2015]	Real Price Net [€2015]	Real Specific price 2, Net [€2015/kW]
Nissan	LEAF ACENTA	2015	80	30	28538	28538	951	356.72	Ford Focus 1.0 EcoBoost	92	17277	20560	17277	188
Nissan	Leaf Visia	2015	80	24	24950	24950	1040	311.87	Ford Focus 1.0 EcoBoost	92	17277	20560	17277	188
Aixam	Mega e-City	2015	4	13	15084	15084	1160	3771.01	Aixam Miniauto L6e	4	6966	8290	6966	1742
Tesla	Model S 60	2015	225	60	54874	54874	915	243.88	BMW 335i	225	37605	44750	37605	167
Peugeot	Partner Electric L1	2015	49	22.5	26100	26100	1160	532.65	Peugeot Partner Kastenwagen	72	14370	17100	14370	200
Kia	Soul EV	2015	81	27	25874	25874	958	319.43	Kia Soul	97	14277	16990	14277	147
German E-Cars	Stromos	2015	56	19.2	28011	28011	1459	500.20	Opel Karl	55	7983	9500	7983	145
Tazzari	Zero	2015	25	15	20471	20471	1365	818.82	Citroën C1	51	7471	8890	7471	146
Mercedes-Benz	B-Klasse Electric Drive	2016	132	28	32900	32818	1172	248.62	Mercedes-Benz B 250	155	27905	33124	27835	180
Citroën	C-Zero	2016	49	14.5	16639	16597	1145	338.72	Citroën C1	51	7639	9067	7620	149
BYD	e6	2016	90	80	50420	50294	629	558.83	Ford Focus 1.0 EcoBoost	92	17739	21057	17695	192
Smart	ED	2016	60	17.6	18437	18391	1045	306.52	Smart fortwo	52	9836	11676	9812	189
Volkswagen	E-Golf	2016	85	24.2	29328	29255	1209	344.17	VW Golf TSI BlueMotion Comfortline	85	19370	22993	19321	227
Nissan	E-NV200 EVALIA	2016	80	24	31248	31170	1299	389.62	Nissan EVALIA	81	18000	21367	17955	222
VW	e-up	2016	60	18.7	22605	22549	1206	375.81	VW up! (move up!)	55	9559	11347	9535	173
Ford	Focus Electric	2016	107	23	29328	29255	1272	273.41	Ford Focus 1.0 EcoBoost	92	17739	21057	17695	192
Hyundai	Hyundai IONIQ Elektro	2016	88	11.5	27983	27913	2427	317.20	Toyota Auris	73	13437	15950	13403	184
Peugeot	i0n	2016	49	14.5	16639	16597	1145	338.72	Citroën C1	51	7639	9067	7620	149
BMW	i3 60 Ah	2016	125	22	29370	29297	1332	234.37	BMW 320i	135	29790	35362	29716	220

Manufacturer	Model	Year	Power [KW]	Battery [kWh]	Net Price [€]	Real Price Net [€₂₀₁₅]	Real Specific Price 1, Net [€₂₀₁₅/kWh]	Real Specific price 2, Net [€₂₀₁₅/kW]	Equivalent ICE Car	Power [KW]	Net Price [€]	Real Price with tax [€₂₀₁₅]	Real Price Net [€₂₀₁₅]	Real Specific price 2, Net [€₂₀₁₅/kW]
BMW	i3 94 Ah	2016	125	33	30378	30302	918	242.42	BMW 320i	135	29790	35362	29716	220
Mitsubishi	i-MiEV	2016	49	16	19992	19942	1246	406.98	Citroën C1	51	7639	9067	7620	149
Nissan	LEAF ACENTA	2016	80	24	27214	27146	1131	339.33	Ford Focus 1.0 EcoBoost	92	17739	21057	17695	192
Nissan	LEAF ACENTA	2016	80	30	28895	28823	961	360.29	Ford Focus 1.0 EcoBoost	92	17739	21057	17695	192
Nissan	LEAF TEKNA	2016	80	24	29231	29158	1215	364.48	Ford Focus 1.0 EcoBoost	92	17739	21057	17695	192
Nissan	LEAF TEKNA	2016	80	30	30912	30835	1028	385.43	Ford Focus 1.0 EcoBoost	92	17739	21057	17695	192
Nissan	Leaf Visia	2016	80	24	24592	24531	1022	306.64	Ford Focus 1.0 EcoBoost	92	17739	21057	17695	192
Nissan	Leaf Visia	2016	80	30	26273	26208	874	327.59	Ford Focus 1.0 EcoBoost	92	17739	21057	17695	192
Aixam	Mega e-City	2016	4	13	13437	13403	1031	3350.87	Aixam City	6	8983	10663	8961	1493
Tesla	Model 90D	2016	310	90	85798	85584	951	276.08	Mercedes-AMG C 63	350	64100	76089	63940	183
Tesla	Model P90D	2016	345	90	104370	104109	1157	301.77	Mercedes-AMG C 63	350	64100	76089	63940	183
Tesla	Model S 60	2016	235	60	64370	64209	1070	273.23	BMW 340i	240	39076	46384	38978	162
Tesla	Model S 60	2016	235	75	72269	72089	961	306.76	BMW 340i	240	39076	46384	38978	162
Tesla	Model X 75D	2016	245	75	89748	89524	1194	365.40	Audi SQ7	320	76471	90773	76280	238
Tesla	Model X 90D	2016	386	90	98908	98661	1096	255.60	Audi SQ7	320	76471	90773	76280	238
Tesla	Model X 100D	2016	568	100	101765	101511	1015	178.72	Audi SQ7	320	76471	90773	76280	238
Peugeot	Partner Electric L1	2016	49	22.5	21290	21237	944	433.40	Peugeot Partner Kastenwagen	72	14350	17033	14314	199
Kia	Soul EV	2016	81	27	25874	25809	956	318.63	Kia Soul	97	14277	16948	14242	147
German E-Cars	Stromos	2016	56	19.2	26471	26405	1375	471.51	Opel Karl	55	7983	9476	7963	145
Tazzari	Zero City	2016	25	8	15300	15262	1908	610.47	Citroën C1	51	7639	9067	7620	149
Tazzari	Zero EM1	2016	30	13	24500	24439	1880	814.63	Citroën C1	51	7639	9067	7620	149

Manufacturer	Model	Year	Power [KW]	Battery [kWh]	Net Price [€]	Real Price Net [€₂₀₁₅]	Real Specific Price 1, Net [€₂₀₁₅/kWh]	Real Specific price 2, Net [€₂₀₁₅/kW]	Equivalent ICE Car	Power [KW]	Net Price [€]	Real Price with tax [€₂₀₁₅]	Real Price Net [€₂₀₁₅]	Real Specific price 2, Net [€₂₀₁₅/kW]
Tazzari	Zero EM2 Space	2016	30	13	21800	21746	1673	724.85	Citroën C1	51	7639	9067	7620	149
Tazzari	Zero Junior	2016	10	5	13800	13766	2753	1376.56	Citroën C1	51	7639	9067	7620	149
Volkswagen	E-Golf	2017	85	24.2	29328	29182	1206	343.315553	VW Golf TSI BlueMotion	85	18782	22239	18688	220
Peugeot	Partner Electric L2	2017	49	22.5	22339	22228	988	453.639878	Comfortline Peugeot Partner Kastenwagen	72	14350	16991	14278	198
Renault	Zoe 41	2017	65	41	27647	27510	671	423.22325	Renault Clio	66	14109	16790	14109	214

Table 13: Price data of PHEVs and their equivalent CVs

Manufacturer	Model	Year	Power ICE [kW]	Power electric engine [kW]	Power [kW]	Battery [kWh]	Net Price [€]	Real Specific Price 1, Net [€₂₀₁₅/kWh]	Real Specific Price 2, Net [€₂₀₁₅/kW]	Equivalent ICE Car	Power [kW]	Net Price [€]	Real Price Net [€₂₀₁₅]	Real Specific Price 2, Net [€₂₀₁₅/kW]
Opel	Ampera	2011	111	63	111	16	36891	2338	337	Opel Astra	132	19147	19419	147
Opel	Ampera	2012	111	63	111	16	38571	2440	352	Opel Astra	132	19147	19380	147
Toyota	Prius III	2012	73	60	100	4.4	30420	6998	308	Toyota Auris	73	13992	14162	194
Opel	Ampera	2013	111	63	111	16	32185	2038	294	Opel Astra	132	19147	19399	147
BMW	i3 60 Ah with range extender	2013	25	125	125	22	33151	1527	269	BMW 320i	135	28067	28437	211
Porsche	Panamera S E-Hybrid	2013	245	70	306	9.4	92781	10000	307	Porsche Panamera 4	243	74381	75360	310
Toyota	Prius III	2013	73	60	100	4.4	30420	7005	308	Toyota Auris	73	13403	13580	186

Manufacturer	Model	Year	Power ICE [kW]	Power electric engine [kW]	Power [kW]	Battery [kWh]	Net Price [€]	Real Specific Price 1, Net [€2015/kWh]	Real Specific Price 2, Net [€2015/kW]	Equivalent ICE Car	Power [kW]	Net Price [€]	Real Price Net [€2015]	Real Specific Price 2, Net [€2015/kW]
Audi	A3 e-tron	2014	110	75	150	8.8	31849	3656	214	Audi A3	132	23277	23512	178
Opel	Ampera	2014	111	63	111	16	32454	2049	295	Opel Astra	125	19761	19960	160
Porsche	Cayenne S E-Hybrid	2014	245	70	306	10.8	68981	6452	228	Porsche Cayenne S	309	67381	68061	220
VW	Golf GTE i3 60 Ah with range extender	2014	110	75	150	8.7	31008	3600	209	VW Golf	110	23193	23428	213
BMW	i8	2014	25	125	125	22	33151	1522	268	BMW 320i	135	28277	28563	212
BMW		2014	170	96	266	5.2	105882	20568	402	Audi r8	316	94958	95917	304
Mitsubishi	Outlander PHEV	2014	89	120	89	12	33605	2829	381	Mitsubishi Outlander 2.0 MIVEC 4WD	110	26588	26857	244
Porsche	Panamera E-Hybrid	2014	245	70	306	9.4	92781	9970	306	Porsche Panamera 4	243	74381	75132	309
Toyota	Prius III	2014	73	60	100	4.4	30420	6984	307	Toyota Auris	73	13403	13539	185
Mercedes-Benz	S 500 e	2014	325	85	325	8.7	91550	10629	285	Mercedes-Benz S 500	335	91550	92475	276
Audi	A3 e-tron	2015	110	75	150	8.8	31849	3619	212	Audi A3	132	22521	22521	171
Mercedes-Benz	C 350 e Kombi	2015	155	60	205	6.4	44150	6899	215	Mercedes-Benz C300 Kombi	180	38125	38125	212
Mercedes-Benz	C 350 e Limousine	2015	155	60	205	6.4	42825	6691	209	Mercedes-Benz C300 Limousine	180	36725	36725	204
Porsche	Cayenne S E-Hybrid	2015	245	70	306	10.8	68981	6387	225	Porsche Cayenne S	309	67381	67381	218
Mercedes-Benz	GLC 350e	2015	155	85	235	8.7	44300	5092	189	Mercedes-Benz GLC 350 d 4MATIC	190	44050	44050	232
Mercedes-Benz	GLE 500 e 4MATIC	2015	245	85	325	8.8	62100	7057	191	Mercedes-Benz GLE 500 4MATIC	335	64600	64600	193

Manufacturer	Model	Year	Power ICE [kW]	Power electric engine [kW]	Power [kW]	Battery [kWh]	Net Price [€]	Real Specific Price 1, Net [€2015/kWh]	Real Specific Price 2, Net [€2015/kW]	Equivalent ICE Car	Power [kW]	Net Price [€]	Real Price Net [€2015]	Real Specific Price 2, Net [€2015/kW]
VW	Golf GTE	2015	110	75	150	8.7	31008	3564	207	VW Golf	110	23193	23193	211
BMW	i3 60 Ah with rage extender	2015	25	125	125	22	33151	1507	265	BMW 320i	135	28782	28782	213
BMW	i8	2015	170	96	266	5.2	105882	20362	398	Audi r8	316	98824	98824	313
Mitsubishi	Outlander PHEV	2015	89	120	89	12	33605	2800	378	Mitsubishi Outlander 2.0 MIVEC 4WD	110	26588	26588	242
Porsche	Panamera E-Hybrid	2015	245	70	306	9.4	92781	9870	303	Panamera 4	243	74381	74381	306
VW	Passat GTE	2015	115	85	160	9.9	37185	3756	232	VW Passat Variant TDI	110	25420	25420	231
VW	Passat Variant GTE	2015	115	85	160	9.9	38025	3841	238	Bluemotion	110	26324	26324	239
Toyota	Prius III	2015	73	60	100	4.4	30420	6914	304	Toyota Auris	73	13437	13437	184
Mercedes-Benz	S 500 e	2015	325	85	325	8.7	92250	10603	284	Mercedes-Benz S 500	335	92250	92250	275
BMW	X5 xDrive40e	2015	180	83	230	9	57479	6387	250	BMW X5 xDrive 35i	225	50336	50336	224
Volvo	XC90 T8 Twin Engine	2015	235	65	300	9.2	64458	7006	215	Volvo XC90 T6 AWD Momentum	235	52714	52714	224
BMW	225xe Active Tourer	2016	100	65	165	7.6	32521	4268	197	BMW 225d	165	33277	33194	201
BMW	330e	2016	135	65	185	7.6	36639	4809	198	BMW 330i	185	33992	33907	183
BMW	740e	2016	190	83	240	9.2	77227	8373	321	BMW 740i	240	74706	74520	310
Audi	A3 e-tron	2016	110	75	150	8.8	33109	3753	220	Audi A3	132	22605	22549	171
Mercedes-Benz	C 350 e Kombi	2016	155	60	205	6.4	44300	6905	216	Mercedes-Benz C300 Kombi	180	38200	38105	212

Manufacturer	Model	Year	Power ICE [kW]	Power electric engine [kW]	Power [kW]	Battery [kWh]	Net Price [€]	Real Specific Price 1, Net [€2015/kWh]	Real Specific Price 2, Net [€2015/kW]	Equivalent ICE Car	Power [kW]	Net Price [€]	Real Price Net [€2015]	Real Specific Price 2, Net [€2015/kW]
Mercedes-Benz	C 350 e Limousine	2016	155	60	205	6.4	42900	6686	209	Mercedes-Benz C300 Limousine	180	36800	36708	204
Porsche	Cayenne S E-Hybrid	2016	245	70	306	10.8	72381	6685	236	Porsche Cayenne S	309	70781	70604	228
Mercedes-Benz	GLC 350e	2016	155	85	235	8.7	44300	5079	188	Mercedes-Benz GLC 350 d 4MATIC	190	44050	43940	231
Mercedes-Benz	GLE 500 e 4MATIC	2016	245	85	325	8.8	62350	7068	191	Mercedes-Benz GLE 500 4MATIC	335	64850	64688	193
VW	Golf GTE	2016	110	75	150	8.7	31008	3555	206	VW Golf	110	21681	21627	197
BMW	i3 60 Ah with Rangeextender	2016	25	125	125	22	33151	1503	265	BMW 320i	135	29790	29716	220
BMW	i3 94 Ah with Rangeextender	2016	25	125	125	33	34160	1033	273	BMW 320i	135	29790	29716	220
BMW	i8	2016	170	96	266	5.2	109244	20956	410	Audi r8	397	139496	139148	350
Kia	Optima Plug-in-Hybrid	2016	115	50	151	9.8	34025	3463	225	Kia Optima	120	21084	21031	175
Mitsubishi	Outlander PHEV	2016	89	120	89	12	33605	2793	377	Mitsubishi MIVEC 4WD	110	26630	26564	241
Porsche	Panamera E-Hybrid	2016	245	70	306	9.4	92781	9846	302	Porsche Panamera 4	243	78181	77986	321
VW	Passat GTE	2016	115	85	160	9.9	37185	3747	232	VW Passat TDI Bluemotion	110	26303	26237	239
VW	Passat Variant GTE	2016	115	85	160	9.9	38025	3831	237	VW Passat Variant TDI Bluemotion	110	28235	28165	256
Toyota	Prius IV	2016	72	53	90	8.8	23655	2681	262	Toyota Auris	73	13437	13403	184

Manufacturer	Model	Year	Power ICE [kW]	Power electric engine [kW]	Power [kW]	Battery [kWh]	Net Price [€]	Real Specific Price 1, Net [€₂₀₁₅/kWh]	Real Specific Price 2, Net [€₂₀₁₅/kW]	Equivalent ICE Car	Power [kW]	Net Price [€]	Real Price Net [€₂₀₁₅]	Real Specific Price 2, Net [€₂₀₁₅/kW]
Audi	Q7 e-tron	2016	190	94	275	17.3	67647	3900	245	Audi Q7	245	54874	54737	223
Mercedes-Benz	S 500 e	2016	325	85	325	8.7	92650	10623	284	Mercedes-Benz S 500	335	92650	92419	276
BMW	xDrive40e X5	2016	180	83	230	9	58655	6501	254	BMW X5 xDrive 35i	225	52101	51971	231
Volvo	XC90 T8 Twin Engine	2016	235	65	300	9.2	64412	6984	214	Volvo XC90 T6 AWD Momentum	235	54496	54360	231
VW	Golf GTE	2017	110	75	150	8.7	31008	3546	206	VW Golf	110	21681	21573	196
Porsche	Panamera 4 E-Hybrid	2017	243	100	340	14.1	90381	6378	265	Panamera 4 Porsche	243	78181	77792	320

Table 14: Monthly fixed and maintenance costs of BEVs and their equivalent CVs

Manufacturer	Model	Monthly Maintenance Costs [EUR/m]	Monthly Fixed Costs [EUR/m]	Equivalent ICE Car	Monthly Maintenance Costs [EUR/m]	Monthly Fixed Costs [EUR/m]
Mercedes-Benz	B-Klasse Electric Drive	56	103	Mercedes-Benz B 250	78	134
German E-Cars	CETOS	49	95	Opel Corsa	52	85
CCS Elektromobile	City-Stromerle Coupe	32	82	Aixam Miniauto L7e	39	75
Citroën	C-Zero	49	95	Citroën C1	39	75
BYD	e6	62	96	Ford Focus 1.0 EcoBoost	68	92
Smart	ED	49	95	Smart fortwo	42	75

Manufacturer	Model	Montly Maintenance Costs [EUR/m]	Monthly Fixed Costs [EUR/m]	Equivalent ICE Car	Montly Maintenance Costs [EUR/m]	Monthly Fixed Costs [EUR/m]
Volkswagen	E-Golf	62	96	VW Golf TSI BlueMotion Comfortline	50	99
Nissan	E-NV200 EVALIA	45	93	Nissan EVALIA	67	108
VW	e-up	32	82	VW up! (move up!)	42	74
Ford	Focus Electric	62	96	Ford Focus 1.0 EcoBoost	68	92
Hyundai	Hyundai IONIQ Elektro	49	95	Toyota Auris	49	91
Peugeot	i0n	66	89	Citroën C1	39	75
BMW	i3 60 Ah	66	89	BMW 320i	73	122
BMW	i3 94 Ah	46	95	BMW 320i	73	122
Mitsubishi	i-MiEV	48	93	Citroën C1	39	75
Fiat	Karabag 500 E	32	82	Fiat 500	66	76
Nissan	LEAF ACENTA	49	105	Ford Focus 1.0 EcoBoost	68	92
Nissan	LEAF TEKNA	56	105	Ford Focus 1.0 EcoBoost	68	92
Nissan	Leaf Visia	49	105	Ford Focus 1.0 EcoBoost	68	92
Aixam	Mega e-City	32	82	Aixam City	39	75
Tesla	Model 90D	173	229	Mercedes-AMG C 63	156	196
Tesla	Model P90D	173	229	Mercedes-AMG C 63	156	196
Tesla	Model S	149	229	BMW 335i	102	154
Tesla	Model S 60	149	229	BMW 340i	102	154
Tesla	Model X 100D	155	172	Audi SQ7	186	232
Tesla	Model X 75D	134	172	Audi SQ7	186	232
Tesla	Model X 90D	155	172	Audi SQ7	186	232
Peugeot	Partner Electric L1	46	95	Peugeot Partner Kastenwagen	68	92
Peugeot	Partner Electric L1	46	95	Peugeot Partner Kastenwagen	68	92
Kia	Soul EV	48	92	Kia Soul	58	104

Manufacturer	Model	Monthly Maintenance Costs [EUR/m]	Monthly Fixed Costs [EUR/m]	Equivalent ICE Car	Monthly Maintenance Costs [EUR/m]	Monthly Fixed Costs [EUR/m]
German E-Cars	Stromos	46	95	Opel Karl	47	94
Tazzari	Zero	32	82	Citroën C1	39	75
Tazzari	Zero	32	82	Citroën C1	39	75
Tazzari	Zero City	32	82	Citroën C1	39	75
Tazzari	Zero EM1	32	82	Citroën C1	39	75
Tazzari	Zero EM2 Space	32	82	Citroën C1	39	75
Tazzari	Zero Junior	32	82	Citroën C1	39	75
Renault	Zoe 41	33	95	Renault Clio	54	96

Table 15: Monthly fixed and maintenance costs of PHEVs and their equivalent CVs

Manufacturer	Model	Monthly Maintenance Costs [EUR/m]	Monthly Fixed Costs [EUR/m]	Equivalent ICE Car	Monthly Maintenance Costs [EUR/m]	Monthly Fixed Costs [EUR/m]
BMW	225xe Active Tourer	71	86	BMW 225d	73	107
BMW	330e	91	116	BMW 330i	91	128
BMW	740e	118	169	BMW 740i	111	181
Audi	A3 e-tron	71	88	Audi A3	62	92
Opel	Ampera	66	93	Opel Astra	67	108
Mercedes-Benz	C 350 e Kombi	107	122	Mercedes-Benz C300 Kombi	99	131
Mercedes-Benz	C 350 e Limousine	107	122	Mercedes-Benz C300 Limousine	99	132
Porsche	Cayenne S E-Hybrid	115	173	Porsche Cayenne S	128	190
Mercedes-Benz	GLC 350e	103	132	Mercedes-Benz GLC 350 d 4MATIC	104	179

Manufacturer	Model	Monthly Maintenance Costs [EUR/m]	Monthly Fixed Costs [EUR/m]	Equivalent ICE Car	Monthly Maintenance Costs [EUR/m]	Monthly Fixed Costs [EUR/m]
Mercedes-Benz	GLE 500 e 4MATIC	156	177	Mercedes-Benz GLE 500 4MATIC	157	195
VW	Golf GTE	66	93	VW Golf	62	87
BMW	i3 60 Ah with Rang-eextender	71	89	BMW 320i	73	122
BMW	i3 94 Ah with Rang-eextender	71	89	BMW 320i	73	122
BMW	i8	124	175	Audi r8	175	257
Kia	Optima Plug-in-Hybrid	80	118	Kia Optima	71	131
Mitsubishi	Outlander PHEV	87	109	Mitsubishi Outlander 2.0 MIVEC 4WD	83	117
Porsche	Panamera 4 E-Hybrid	181	160	Porsche Panamera 4	145	173
Porsche	Panamera E-Hybrid	181	160	Porsche Panamera 4	145	173
Porsche	Panamera S E-Hybrid	181	160	Porsche Panamera 4	145	173
VW	Passat GTE	78	88	VW Passat TDI Bluemotion	58	123
VW	Passat Variant GTE	78	88	VW Passat Variant TDI Bluemotion	59	92
Toyota	Prius III	53	118	Toyota Auris	49	91
Toyota	Prius IV	53	118	Toyota Auris	49	91
Audi	Q7 e-tron	118	176	Audi Q7	95	167
Mercedes-Benz	S 500 e	171	171	Mercedes-Benz S 500	164	190
BMW	X5 xDrive40e	79	166	BMW X5 xDrive 35i	100	194
Volvo	XC90 T8 Twin Engine	124	143	Volvo XC90 T6 AWD Momentum	112	158

Table 16: Energy consumption and CO_2 emissions of BEVs and their conventional CVs

Manufacturer	Model	Energy Consumption [kWh/100 km]	Equivalent car	Energy Consumption [l/100 km]	CO_2-Emissions [g/km]
Mercedes-Benz	B-Klasse Electric Drive	16.6	Mercedes-Benz B 250	6.7	155
German E-Cars	CETOS	17.5	Opel Corsa	5.3	124
CCS Elektromobile	City-Stromerie Coupe	10	Aixam Miniauto L7e	6	143
Citroën	C-Zero	12.6	Citroën C1	4.3	99
BYD	e6	21.6	Ford Focus 1.0 EcoBoost	4.7	108
Smart	ED	15.1	Smart fortwo	4.1	93
Volkswagen	E-Golf	12.7	VW Golf TSI BlueMotion Comfortline	4.3	99
Nissan	E-NV200 EVALIA	16.5	Nissan EVALIA	4.9	133
VW	e-up	11.7	VW up! (move up!)	4.6	106
Ford	Focus Electric	15.4	Ford Focus 1.0 EcoBoost	4.7	108
Hyundai	Hyundai IONIQ Elektro	11.5	Toyota Auris	73	129
Peugeot	iOn	12.6	Citroën C1	4.1	95
BMW	i3 60 Ah	12.6	BMW 320i	6.3	147
BMW	i3 94 Ah	12.6	BMW 320i	5.3	124
Mitsubishi	i-MiEV	12.5	Citroën C1	4.1	95
Fiat	Karabag 500 E	11	Fiat 500	4.8	113
Nissan	LEAF ACENTA	15	Ford Focus 1.0 EcoBoost	4.7	108
Nissan	LEAF TEKNA	15	Ford Focus 1.0 EcoBoost	4.7	108
Nissan	LeafVisia	15	Ford Focus 1.0 EcoBoost	4.7	108
Aixam	Mega e-City	18.6	Aixam / Ligier	3.6	86
Aixam	Mega e-City	18.6	Aixam Miniauto L6e	2.96	77.9
Aixam	Mega e-City	18.6	Aixam City	3.1	80
Tesla	Model P90D	17.7	Mercedes-AMG C 63	8.2	192
Tesla	Model S 60	17.7	BMW 335i	7.9	186
Tesla	Model S 60	17.7	BMW 340i	7.5	174
Tesla	Model X 90D	19.7	Audi SQ7	7.4	194

Manufacturer	Model	Energy Consumption [kWh/100 km]	Equivalent car	Energy Consumption [l/100 km]	CO_2-Emissions [g/km]
Peugeot	Partner Electric L2	17.7	Peugeot Partner Kastenwagen	6.5	150
Kia	Soul EV	14.7	Kia Soul	6.8	158
German E-Cars	Stromos	17.5	Opel Agila B	4.7	109
German E-Cars	Stromos	17.5	Opel Karl	4.4	102
Tazzari	Zero	13	Citroën C1	4.1	95
Tazzari	Zero City	13	Citroën C1	4.1	95
Tazzari	Zero EM1	13	Citroën C1	4.1	95
Tazzari	Zero EM2 Space	13	Citroën C1	4.1	95
Tazzari	Zero Junior	13	Citroën C1	4.1	95
Renault	Zoe 41	10.25	Renault Clio	5.6	128.5

Table 17: Energy consumption and CO_2 emissions of PHEVs and their conventional CVs

Manufacturer	Model	Energy Consumption [kWh/100 km]	Energy Consumption [l/100 km]	CO_2-Emissions CE [g/km]	Equivalent car	Energy Consumption [l/100 km]	CO_2-Emissions [g/km]
BMW	225xe Active Tourer	11.8	2	46	BMW 225d	4.45	117.5
BMW	330e	11.45	2	46.5	BMW 330i	6.3	147
BMW	740e	12.9	2.15	49.5	BMW 740i	6.9	161
Audi	A3 e-tron	11.5	1.7	38	Audi A3	5.8	135
Opel	Ampera	13.5	1.2	27	Opel Astra	6.8	159
Opel	Ampera	13.5	1.2	27	Opel Astra	6.1	152
Mercedes-Benz	C 350 e Kombi	11.7	2.25	52	Mercedes-Benz C300 Kombi	6.55	148.5
Mercedes-Benz	C 350 e Limousine	11.7	2.25	52	Mercedes-Benz C300 Limousine	6.55	148.5
Porsche	Cayenne S E-Hybrid	19.7	3.35	77	Porsche Cayenne S	9.65	226

Manufacturer	Model	Energy Consumption [kWh/100 km]	Energy Consumption [l/100 km]	CO_2-Emissions CE [g/km]	Equivalent car		CO_2-Emissions [g/km]	Energy Consumption [l/100 km]	CO_2-Emissions [g/km]
Mercedes-Benz	GLC 350e	14.65	2.6	61.5	Mercedes-Benz GLC 350 d 4MATIC		6.05	164
Mercedes-Benz	GLE 500 e 4MATIC	17.75	3.5	81	Mercedes-Benz GLE 500 4MATIC		10.7	245
VW	Golf GTE	11.9	1.6	42	VW Golf		5.25	120.5
BMW	i3 60 Ah with Rangeextender	13.5	0.6	13	BMW 320i		6.4	147.5
BMW	i3 94 Ah with Rangeextender	13.5	0.6	13	BMW 320i		6.4	147.5
BMW	i8	11.9	2.1	49	Audi r8		14.2	332
Kia	Optima Plug-in-Hybrid	12.2	1.6	37	Kia Optima		7.4	173
Mitsubishi	Outlander PHEV	13.4	1.8	42	Mitsubishi Outlander 2.0 MIVEC 4WD		6.7	155
Porsche	Panamera 4 E-Hybrid	15.9	2.5	56	Porsche Panamera 4		7.75	176
Porsche	Panamera S E-Hybrid	16.2	3.1	71	Porsche Panamera 4		8.7	203
VW	Passat GTE	12.8	1.75	39	VW Passat TDI Bluemotion		4.25	110.5
VW	Passat Variant GTE	12.95	1.75	39	VW Passat Variant TDI Bluemotion		4.25	111.5
Toyota	Prius III	7.2	1	89	Toyota Auris		5.55	129
Toyota	Prius IV	7.2	1	73	Toyota Auris		5.55	129
Audi	Q7 e-tron	18.55	1.8	48	Audi Q7		10.7	249
Mercedes-Benz	S 500 e	15.5	2.8	65	Mercedes-Benz S 500		8.55	199
BMW	X5 xDrive40e	15.35	3.35	77.5	BMW X5 xDrive 35i		8.5	198
Volvo	XC90 T8 Twin Engine	18.2	2.1	49	Volvo XC90 T6 AWD Momentum		7.7	179

Table 18: Real users costs of BEVs and their equivalent CVs in the specific years

Manufacturer	Model	Year	Real Price [€/km]	Real Price On Road [€/km]	Equivalent car	Real Price [€/km]	Real Price Diesel	Real Price On Road [€/km]
Mitsubishi	i-MiEV	2010	0.57	0.58	Citroën C1	0.26	0.20	0.28

Manufacturer	Model	Year	Real Price [€/km]	Real Price On Road [€/km]	Equivalent car	Real Price [€/km]	Real Price Diesel [€/km]	Real Price On Road [€/km]
Fiat	Karabag 500 E	2010	0.85	0.86	Fiat 500	0.34	0.26	0.35
German E-Cars	Stromos	2010	0.67	0.68	Opel Agila B / Suzuki Splash 2010	0.31	0.24	0.33
Tazzari	Zero	2010	0.42	0.43	Citroën C1	0.26	0.20	0.28
Mitsubishi	i-MiEV	2011	0.57	0.58	Citroën C1	0.26	0.20	0.28
Fiat	Karabag 500 E	2011	0.63	0.64	Fiat 500	0.34	0.27	0.36
Aixam	Mega e-City	2011	0.48	0.49	Aixam / Ligier	0.27	0.21	0.28
German E-Cars	Stromos	2011	0.66	0.68	Opel Agila B / Suzuki Splash	0.31	0.24	0.33
Tazzari	Zero	2011	0.42	0.43	Citroën C1	0.26	0.20	0.28
German E-Cars	CETOS	2012	0.69	0.70	Opel Corsa	0.33	0.25	0.36
Citroën	C-Zero	2012	0.50	0.51	Citroën C1	0.26	0.19	0.28
Smart	ED	2012	0.44	0.45	Smart fortwo	0.29	0.23	0.31
Peugeot	iOn	2012	0.50	0.51	Citroën C1	0.26	0.19	0.28
Mitsubishi	i-MiEV	2012	0.64	0.65	Citroën C1	0.26	0.19	0.28
Fiat	Karabag 500 E	2012	0.60	0.61	Fiat 500	0.33	0.26	0.35
Nissan	Leaf Visia	2012	0.53	0.54	Ford Focus 1.0 EcoBoost	0.45	0.37	0.47
German E-Cars	Stromos	2012	0.43	0.44	Opel Agila B / Suzuki Splash	0.31	0.24	0.33
Tazzari	Zero	2012	0.67	0.68	Citroën C1	0.26	0.19	0.28
German E-Cars	CETOS	2013	0.52	0.53	Opel Corsa	0.33	0.26	0.36
Citroën	C-Zero	2013	0.44	0.45	Citroën C1	0.26	0.20	0.28
Smart	ED	2013	0.45	0.47	Smart fortwo	0.29	0.23	0.31
VW	e-up	2013	0.64	0.65	VW up! (cheer up)	0.31	0.24	0.34
Ford	Focus Electric	2013	0.51	0.52	Ford Focus 1.0 EcoBoost	0.45	0.38	0.48
Peugeot	iOn	2013	0.58	0.59	Citroën C1	0.26	0.20	0.28
BMW	i3 60 Ah	2013	0.50	0.51	BMW 320i	0.65	0.56	0.68
Mitsubishi	i-MiEV	2013	0.64	0.65	Citroën C1	0.26	0.20	0.28
Fiat	Karabag 500 E	2013	0.51	0.52	Fiat 500	0.33	0.26	0.35
Nissan	Leaf Visia	2013	1.33	1.34	Ford Focus 1.0 EcoBoost	0.45	0.38	0.48

Manufacturer	Model	Year	Real Price [€/km]	Real Price On Road [€/km]	Equivalent car	Real Price [€/km]	Real Price Diesel	Real Price On Road [€/km]
Tesla	Model S	2013	1.21	1.22	BMW 335i	0.85	0.73	0.89
Tesla	Model S 60	2013	0.54	0.55	BMW 335i	0.85	0.73	0.89
German E-Cars	Stromos	2013	0.43	0.44	Opel Agila B / Suzuki Splash	0.31	0.24	0.34
Tazzari	Zero	2013	0.63	0.64	Citroën C1	0.26	0.20	0.28
Mercedes-Benz	B-Klasse Electric Drive	2014	0.68	0.70	Mercedes-Benz B 250	0.69	0.59	0.73
German E-Cars	CETOS	2014	0.51	0.53	Opel Corsa	0.33	0.26	0.37
Citroën	C-Zero	2014	0.43	0.44	Citroën C1	0.26	0.20	0.29
Smart	ED	2014	0.59	0.60	Smart fortwo	0.29	0.23	0.31
Volkswagen	E-Golf	2014	0.45	0.46	VW Golf TSI BlueMotion Comfortline	0.46	0.40	0.49
VW	e-up	2014	0.64	0.65	VW up! (cheer up)	0.31	0.24	0.34
Ford	Focus Electric	2014	0.51	0.52	Ford Focus 1.0 EcoBoost	0.45	0.38	0.48
Peugeot	iOn	2014	0.58	0.59	Citroën C1	0.26	0.20	0.29
BMW	i3 60 Ah	2014	0.43	0.44	BMW 320i	0.65	0.56	0.69
Mitsubishi	i-MiEV	2014	0.45	0.46	Citroën C1	0.26	0.20	0.29
Fiat	Karabag 500 E	2014	0.51	0.52	Fiat 500	0.33	0.26	0.36
Nissan	Leaf Visia	2014	1.13	1.14	Ford Focus 1.0 EcoBoost	0.45	0.38	0.48
Tesla	Model S 60	2014	0.53	0.55	BMW 335i	0.85	0.74	0.90
Peugeot	Partner Electric L1	2014	0.53	0.54	Peugeot Partner Kastenwagen	0.44	0.35	0.48
Kia	Soul EV	2014	0.53	0.54	Kia Soul	0.44	0.34	0.48
German E-Cars	Stromos	2014	0.43	0.44	Opel Agila B	0.31	0.24	0.34
Tazzari	Zero	2014	0.63	0.64	Citroën C1	0.26	0.20	0.29
Mercedes-Benz	B-Klasse Electric Drive	2015	0.28	0.29	Mercedes-Benz B 250	0.68	0.59	0.73
CCS Elektromobile	City-Stromerle Coupe	2015	0.49	0.50	Aixam Miniauto L7e	0.34	0.25	0.37
Citroën	C-Zero	2015	0.44	0.45	Citroën C1	0.26	0.20	0.29
Smart	ED	2015	0.58	0.59	Smart fortwo	0.29	0.23	0.31

Manufacturer	Model	Year	Real Price [€/km]	Real Price On Road [€/km]	Equivalent car	Real Price [€/km]	Real Price Diesel	Real Price On Road [€/km]
Volkswagen	E-Golf	2015	0.44	0.45	VW Golf TSI BlueMotion Comfortline	0.46	0.39	0.48
VW	e-up	2015	0.57	0.58	VW up! (cheer up)	0.31	0.24	0.34
Ford	Focus Electric	2015	0.37	0.38	Ford Focus 1.0 EcoBoost	0.45	0.37	0.48
Peugeot	iOn	2015	0.57	0.58	Citroën C1	0.26	0.20	0.29
BMW	i3 60 Ah	2015	0.43	0.44	BMW 320i	0.66	0.56	0.70
Mitsubishi	i-MiEV	2015	0.56	0.57	Citroën C1	0.26	0.20	0.29
Nissan	LEAF ACENTA	2015	0.57	0.58	Ford Focus 1.0 EcoBoost	0.45	0.37	0.48
Nissan	LEAF ACENTA	2015	0.52	0.53	Ford Focus 1.0 EcoBoost	0.45	0.37	0.48
Nissan	LeafVisia	2015	0.35	0.36	Ford Focus 1.0 EcoBoost	0.45	0.37	0.48
Aixam	Mega e-City	2015	1.13	1.15	Aixam Miniauto L6e	0.24	0.19	0.25
Tesla	Model S 60	2015	0.53	0.54	BMW 335i	0.85	0.74	0.90
Peugeot	Partner Electric L1	2015	0.53	0.54	Peugeot Partner Kastenwagen	0.43	0.33	0.47
Kia	Soul EV	2015	0.55	0.56	Kia Soul	0.44	0.33	0.48
German E-Cars	Stromos	2015	0.43	0.44	Opel Karl	0.29	0.23	0.32
Tazzari	Zero	2015	0.62	0.64	Citroën C1	0.26	0.20	0.29
Mercedes-Benz	B-Klasse Electric Drive	2016	0.40	0.41	Mercedes-Benz B 250	0.66	0.56	0.71
Citroën	C-Zero	2016	0.87	0.88	Citroën C1	0.26	0.20	0.29
BYD	e6	2016	0.43	0.45	Ford Focus 1.0 EcoBoost	0.45	0.38	0.48
Smart	ED	2016	0.58	0.59	Smart fortwo	0.30	0.23	0.32
Volkswagen	E-Golf	2016	0.58	0.59	VW Golf TSI BlueMotion Comfortline	0.46	0.39	0.48
Nissan	E-NV200 EVALIA	2016	0.45	0.47	Nissan EVALIA	0.47	0.40	0.50
VW	e-up	2016	0.57	0.58	VW up! (move up!)	0.30	0.23	0.33
Ford	Focus Electric	2016	0.55	0.56	Ford Focus 1.0 EcoBoost	0.45	0.38	0.48
Hyundai	Hyundai IONIQ Elektro	2016	0.39	0.40	Toyota Auris	0.41	0.30	0.46
Peugeot	iOn	2016	0.57	0.58	Citroën C1	0.26	0.20	0.29

Manufacturer	Model	Year	Real Price [€/km]	Real Price On Road [€/km]	Equivalent car	Real Price [€/km]	Real Price Diesel [€/km]	Real Price On Road [€/km]
BMW	i3 60 Ah	2016	0.57	0.58	BMW 320i	0.67	0.58	0.71
BMW	i3 94 Ah	2016	0.43	0.44	BMW 320i	0.66	0.58	0.69
Mitsubishi	i-MiEV	2016	0.54	0.55	Citroën C1	0.26	0.20	0.29
Nissan	LEAF ACENTA	2016	0.57	0.58	Ford Focus 1.0 EcoBoost	0.45	0.38	0.48
Nissan	LEAF ACENTA	2016	0.58	0.59	Ford Focus 1.0 EcoBoost	0.45	0.38	0.48
Nissan	LEAF TEKNA	2016	0.60	0.62	Ford Focus 1.0 EcoBoost	0.45	0.38	0.48
Nissan	LEAF TEKNA	2016	0.51	0.52	Ford Focus 1.0 EcoBoost	0.45	0.38	0.48
Nissan	Leaf Visia	2016	0.53	0.55	Ford Focus 1.0 EcoBoost	0.45	0.38	0.48
Nissan	Leaf Visia	2016	0.32	0.33	Ford Focus 1.0 EcoBoost	0.45	0.38	0.48
Aixam	Mega e-City	2016	1.58	1.59	Aixam City	0.27	0.22	0.29
Tesla	Model 90D	2016	1.83	1.85	Mercedes-AMG C 63	1.31	1.18	1.36
Tesla	Model P90D	2016	1.26	1.27	Mercedes-AMG C 63	1.31	1.18	1.36
Tesla	Model S 60	2016	1.37	1.38	BMW 340i	0.87	0.76	0.91
Tesla	Model S 60	2016	1.55	1.56	BMW 340i	0.87	0.76	0.91
Tesla	Model X 75D	2016	1.70	1.71	Audi SQ7	1.52	1.41	1.57
Tesla	Model X 90D	2016	1.74	1.76	Audi SQ7	1.52	1.41	1.57
Tesla	Model X 100D	2016	0.47	0.49	Audi SQ7	1.52	1.41	1.57
Peugeot	Partner Electric L1	2016	0.52	0.54	Peugeot Partner Kastenwagen	0.43	0.33	0.47
Kia	Soul EV	2016	0.53	0.54	Kia Soul	0.43	0.33	0.48
German E-Cars	Stromos	2016	0.36	0.37	Opel Karl	0.29	0.23	0.32
Tazzari	Zero City	2016	0.47	0.48	Citroën C1	0.26	0.20	0.29
Tazzari	Zero EM1	2016	0.43	0.44	Citroën C1	0.26	0.20	0.29
Tazzari	Zero EM2 Space	2016	0.32	0.33	Citroën C1	0.26	0.20	0.29
Tazzari	Zero Junior	2016	0.57	0.58	Citroën C1	0.26	0.20	0.29

Table 19: Real user costs of PHEVs and their equivalent CVs in the specific years

Manufacturer	Model	Year	Real Price [€/km]	Real Price On Road [€/km]	Equivalent car	Real Price [€/km]	Real Price Diesel [€/km]	Real Price On Road [€/km]
Opel	Ampera	2011	0.71	0.76	Opel Astra	0.52	0.42	0.55
Opel	Ampera	2012	0.73	0.78	Opel Astra	0.52	0.42	0.55
Toyota	Prius III	2012	0.61	0.64	Toyota Auris	0.40	0.31	0.42
Opel	Ampera	2013	0.64	0.69	Opel Astra	0.52	0.42	0.55
BMW	i3 60 Ah with range extender	2013	1.64	1.67	BMW 320i	1.41	1.31	1.44
Porsche	Panamera S E-Hybrid	2013	0.66	0.77	Porsche Panamera 4	0.44	0.31	0.48
Toyota	Prius III	2013	0.64	0.67	Toyota Auris	0.64	0.56	0.67
Audi	A3 e-tron	2014	0.64	0.70	Audi A3	0.54	0.46	0.58
Opel	Ampera	2014	0.64	0.69	Opel Astra	0.51	0.42	0.55
Porsche	Cayenne S E-Hybrid	2014	1.31	1.44	Porsche Cayenne S	1.36	1.21	1.41
VW	Golf GTE	2014	0.62	0.69	VW Golf	0.53	0.45	0.56
BMW	i3 60 Ah with range extender	2014	1.78	1.81	BMW 320i	1.79	1.70	1.83
BMW	i8	2014	0.70	0.78	Audi r8	0.75	0.54	0.84
Mitsubishi	Outlander PHEV	2014	1.65	1.72	Mitsubishi Outlander 2.0 MIVEC 4WD	1.41	1.31	1.45
Porsche	Panamera E-Hybrid	2014	0.66	0.77	Porsche Panamera 4	0.43	0.31	0.49
Toyota	Prius III	2014	1.65	1.78	Toyota Auris	1.66	1.58	1.70
Mercedes-Benz	S 500 e	2014	0.68	0.79	Mercedes-Benz S 500	0.69	0.56	0.74
Audi	A3 e-tron	2015	0.63	0.70	Audi A3	0.53	0.44	0.56
Mercedes-Benz	C 350 e Kombi	2015	0.87	0.95	Mercedes-Benz C300 Kombi	0.82	0.72	0.86
Mercedes-Benz	C 350 e Limousine	2015	0.85	0.93	Mercedes-Benz C300 Limousine	0.80	0.70	0.84

Manufacturer	Model	Year	Real Price [€/km]	Real Price On Road [€/km]	Equivalent car	Real Price [€/km]	Real Price Diesel [€/km]	Real Price On Road [€/km]
Porsche	Cayenne S E-Hybrid	2015	1.30	1.43	Porsche Cayenne S	1.35	1.20	1.41
Mercedes-Benz	GLC 350e	2015	0.89	0.99	Mercedes-Benz GLC 350 d 4MATIC	0.94	0.85	0.98
Mercedes-Benz	GLE 500 e 4MATIC	2015	1.24	1.37	Mercedes-Benz GLE 500 4MATIC	1.35	1.19	1.42
VW	Golf GTE	2015	0.62	0.68	VW Golf	0.53	0.45	0.56
BMW	i3 60 Ah with rage extender	2015	1.77	1.80	BMW 320i	1.83	1.74	1.87
BMW	i8	2015	0.69	0.77	Audi r8	0.75	0.54	0.84
Mitsubishi	Outlander PHEV	2015	1.64	1.71	Mitsubishi Outlander 2.0 MIVEC 4WD	1.40	1.30	1.44
Porsche	Panamera E-Hybrid	2015	0.75	0.86	Porsche Panamera 4	0.63	0.51	0.69
VW	Passat GTE	2015	0.73	0.80	VW Passat TDI Bluemotion	0.56	0.49	0.58
VW	Passat Variant GTE	2015	0.63	0.69	VW Passat Variant TDI Bluemotion	0.37	0.30	0.39
Toyota	Prius III	2015	1.60	1.64	Toyota Auris	1.66	1.58	1.70
Mercedes-Benz	S 500 e	2015	1.09	1.19	Mercedes-Benz S 500	1.07	0.95	1.13
BMW	X5 xDrive40e	2015	1.21	1.33	BMW X5 xDrive 35i	1.09	0.96	1.14
Volvo	XC90 T8 Twin Engine	2015	0.68	0.76	Volvo XC90 T6 AWD Momentum	0.68	0.56	0.73
BMW	225xe Active Tourer	2016	0.64	0.72	BMW 225d	0.68	0.61	0.77
BMW	330e	2016	0.74	0.82	BMW 330i	0.75	0.66	0.75
BMW	740e	2016	1.38	1.46	BMW 740i	1.38	1.28	1.41
Audi	A3 e-tron	2016	0.65	0.71	Audi A3	0.53	0.44	0.59
Mercedes-Benz	C 350 e Kombi	2016	0.87	0.95	Mercedes-Benz C300 Kombi	0.82	0.72	0.84
Mercedes-Benz	C 350 e Limousine	2016	0.85	0.93	Mercedes-Benz C300 Limousine	0.80	0.70	0.84
Porsche	Cayenne S E-Hybrid	2016	1.35	1.47	Porsche Cayenne S	1.39	1.25	1.39
Mercedes-Benz	GLC 350e	2016	0.89	0.99	Mercedes-Benz GLC 350 d 4MATIC	0.94	0.85	1.05
Mercedes-Benz	GLE 500 e 4MATIC	2016	1.24	1.37	Mercedes-Benz GLE 500 4MATIC	1.35	1.19	1.32

Manufacturer	Model	Year	Real Price [€/km]	Real Price On Road [€/km]	Equivalent car	Real Price [€/km]	Real Price Diesel [€/km]	Real Price On Road [€/km]
VW	Golf GTE	2016	0.62	0.68	VW Golf	0.50	0.43	0.65
BMW	i3 60 Ah with Rangeextender	2016	1.81	1.84	BMW 320i	2.39	2.30	2.41
BMW	i3 94 Ah with Rangeextender	2016	0.68	0.71	BMW 320i	0.56	0.46	0.60
BMW	i8	2016	0.69	0.77	Audi r8	0.75	0.54	0.67
Kia	Optima Plug-in-Hybrid	2016	1.63	1.69	Kia Optima	1.46	1.35	1.65
Mitsubishi	Outlander PHEV	2016	0.72	0.79	Mitsubishi Outlander 2.0 MIVEC 4WD	0.62	0.52	0.67
Porsche	Panamera E-Hybrid	2016	0.76	0.87	Porsche Panamera 4	0.65	0.52	0.66
VW	Passat GTE	2016	0.53	0.60	VW Passat TDI Bluemotion	0.37	0.30	0.49
VW	Passat Variant GTE	2016	1.25	1.31	VW Passat Variant TDI Bluemotion	1.04	0.98	1.07
Toyota	Prius IV	2016	1.61	1.64	Toyota Auris	1.66	1.58	1.67
Audi	Q7 e-tron	2016	1.10	1.17	Audi Q7	1.13	0.97	1.09
Mercedes-Benz	S 500 e	2016	1.20	1.30	Mercedes-Benz S 500	1.11	0.98	1.21
BMW	X5 xDrive40e	2016	0.69	0.81	BMW X5 xDrive 35i	0.70	0.58	0.76
Volvo	XC90 T8 Twin Engine	2016	0.69	0.77	Volvo XC90 T6 AWD Momentum	0.69	0.58	0.76
VW	Golf GTE	2017	0.62	0.68	VW Golf	0.50	0.42	0.42
Porsche	Panamera 4 E-Hybrid	2017	1.62	1.71	Porsche Panamera 4	1.46	1.35	1.35

Appendix B: Vehicle price as function of rated engine power and battery capacity

We conduct a linear regression analysis to assess whether rated power [kW] and battery capacity [kWh] (in the case of BEVs) represent suitable functional units to normalize absolute vehicle prices. We conduct the analysis separately for BEVs, PHEVs, and CVs. Due to the sometimes low number of data points in individual years, we decided to include in the regression analysis all vehicle prices identified for the period between 2010 and 2016. As particularly BEVs and PHEVs show substantial technological learning, the ratio between absolute price and rated power of these vehicles arguably declines over the years. We would therefore argue that the coefficients of determination identified here may reflect a "worst case" scenario and could be higher, if data for individual years had been considered.

With this in mind, we observe that the real absolute price of BEVs, PHEVs and CVs indeed increases linearly with the rated engine power (Figure 15Figure 17). Within the value range considered here, the average absolute vehicle price rises by around 192 EUR for BEVs, 260 EUR for PHEVs and 276 EUR for CVs with each additional kilowatt rated engine power. The high coefficients of determination, i.e., 0.80 for BEVs, 0.74 for PHEVs and 0.88 for CVs indeed suggest a robust linear relationship between absolute vehicle price and rated engine power (Figure 15 Figure 17).

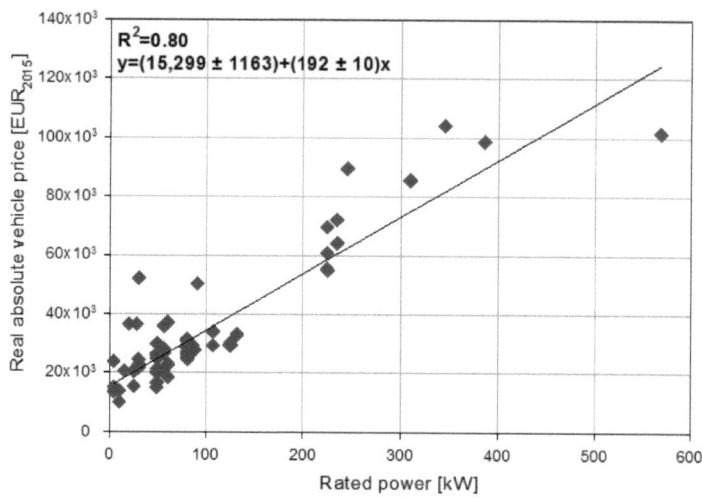

Figure 15: Real absolute price of BEVs as function of rated engine power

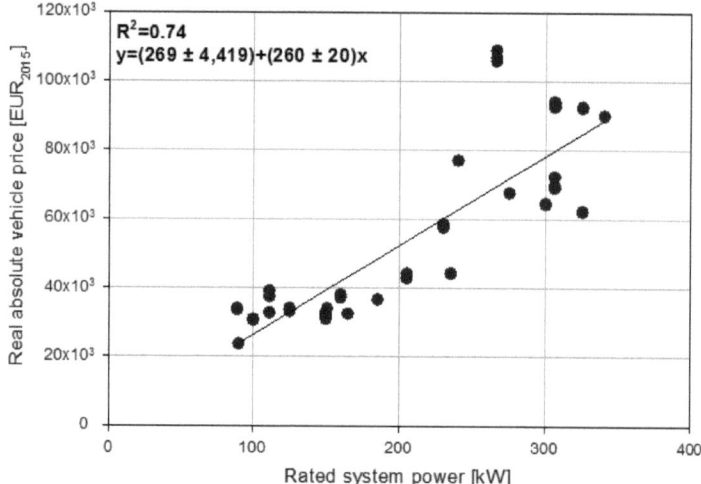

Figure 16: Real absolute price of PHEVs as function of rated combined power output of electric engine and combustion engine

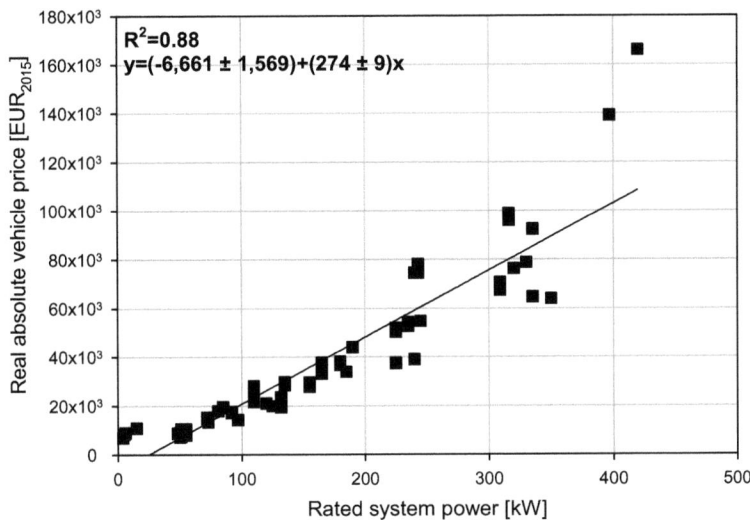

Figure 17: Real absolute price of CVs as function of rated engine power

For BEVs, we also find a positive correlation between the real absolute price and the battery capacity. Within the value range considered here, the real absolute price increases by around 828 EUR with each kilowatt-hour increase in battery capacity. The coefficient of determination of around 0.83 suggests indeed a robust linear relationship between the two parameters (Figure 18).

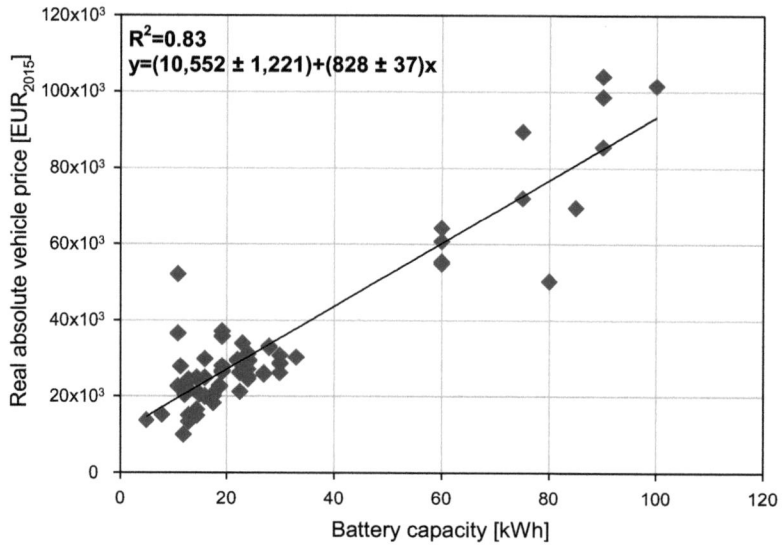

Figure 18: Real absolute price of BEVs as function of battery capacity

For PHEVs, by contrast, absolute price and battery capacity are not correlated (see Figure 19), most likely because of a heterogeneity in the PHEV-data that include both parallel and series PHEVs. The two PHEV types differ considerably in the size and power output of the electric and combustion engines.

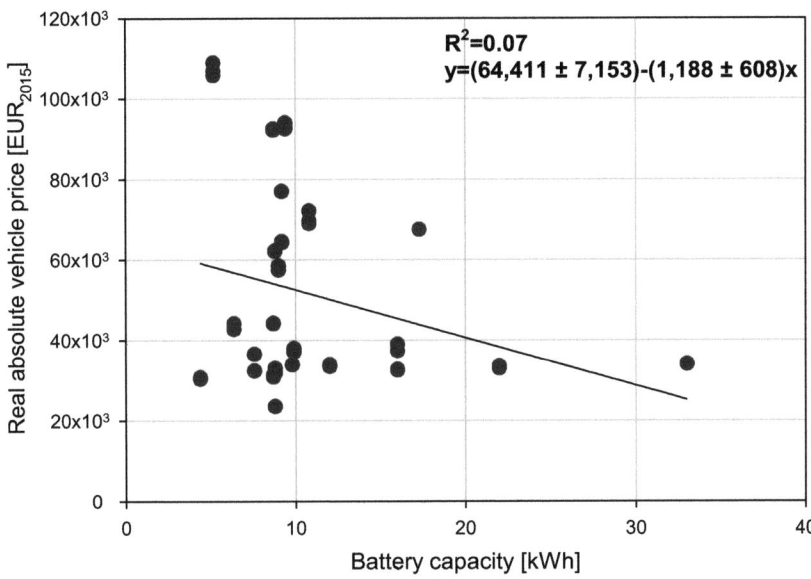

Figure 19: Real absolute price of PHEVs as function of battery capacity

Consistent with these results, we find a linear relationship between the battery capacity and engine power output for BEVs but not so for PHEVs (Figure 20). For BEVs, the battery capacity rises by around 0.21 kWh with each kilowatt increase in power output. The coefficient of determination of around 0.82 suggests a robust linear relationship (see Figure 20).

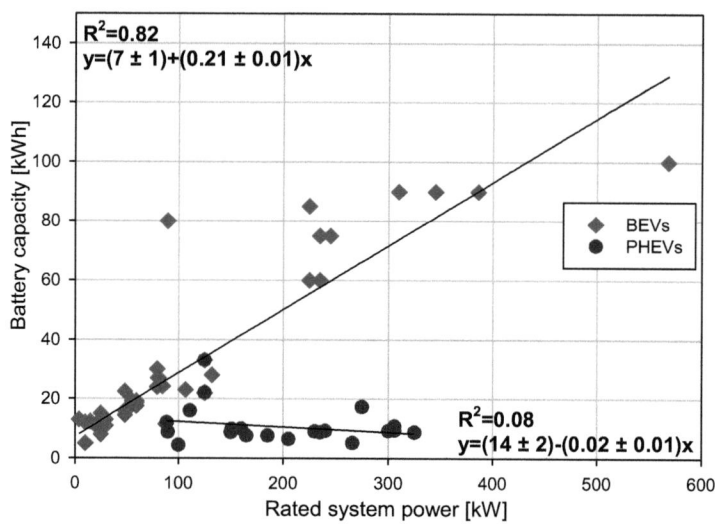

Figure 20: Battery capacity as a function of electric engine power output (BEV) and system power output (PHEV)

This observation suggests that in the case of BEVs, both rated power and battery capacity can be used synonymously as functional unit to normalize absolute vehicle prices. Based on the findings of this analysis, we regard it justified to (i) use the rated power as functional unit for the analysis of vehicle price and (ii) conduct complementary experience curve analyses for BEVs based on vehicle prices normalized by rated engine power and battery capacity.

Appendix C: Time series of real absolute vehicle price and rated engine power

We find increasing net purchasing prices for CVs by 367% for the period under consideration. In contrast, prices for BEVs and PHEVs increase less (15% and 44%), see Figure 21.

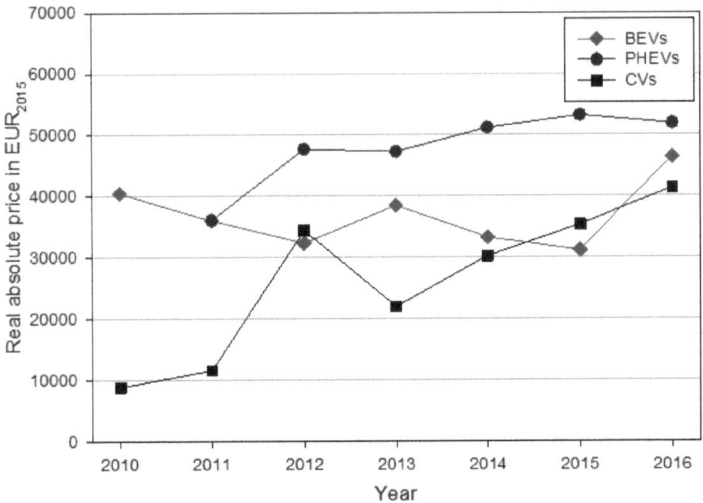

Figure 21: Mean absolute prices of BEVs, PHEVs and CVs from 2010 to 2016

We also find an increase in the average engine power output for the period under consideration, meaning a shift towards more luxury and powerful vehicles contradict the price decrease due to technological learning. The average engine power output of equivalent vehicles raised by 264% from around 50 kW in 2010 to 181 kW in 2016. Considering BEVs, it raised by 238% from around 36 kW in 2010 to 127 kW in 2016. For PHEVs we find an increase by 87% from around 110 kW in 2011 to 207 kW in 2016, see Figure 22.

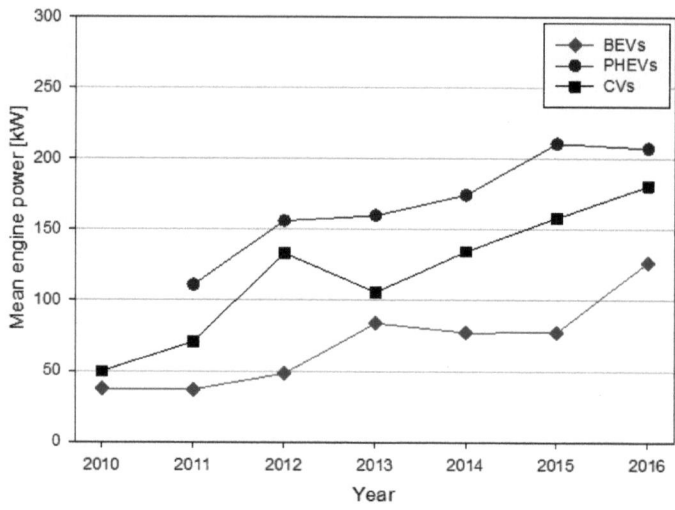

Figure 22: Mean engine power of BEVs, PHEVs, CVs from 2010 to 2016; data for PHEVs represent the maximum rated system power of the electric engine and the internal combustion engine combined

Appendix D: Benchmarking the costs of mitigating NO$_X$ and PN emissions by BEVs and PHEVs

The costs of mitigating NO$_X$ and PN emissions through the deployment of BEVs and PHEVs can be compared to the costs of mitigation technologies applied in light-duty and heavy-duty vehicles as well as in the manufacturing industry. Table 20 and Table 21 provide an overview of selected technologies, data sources, assumptions used, and mitigation costs calculated. Values are ether obtained from external sources or calculated as:

$$C_m = \frac{C_t}{E * M_{Li}}$$

Where C_m [EUR/t NO$_X$; EUR/10^{17} particles] represents the mitigation costs of a technology, C_t the costs of the respective technology [EUR], E the efficacy, meaning the mitigated emissions due to the used respective technology [g NOx/km; particles/km] and M_{Li} the total mileage throughout the assumed lifetime of the vehicle [km].

In view of controlling the emissions of conventional vehicles, it is noteworthy that vehicle manufacturers are required by law to ensure proper functioning of after-treatment technologies throughout the useful vehicle life, e.g., for up to 5 years or 160,000 km in the case of passenger cars (EC, 2007). Yet, aging tends to decrease the efficacy of after-treatment technologies (within and beyond the 5 year period) in an often case specific manner that is difficult to quantify. We therefore assume here that the efficacy assumed in Table 20 and Table 21 remains constant throughout the assumed life time of after-treatment systems. This simplification is justified given the absence of reliable data but might underestimate the actual costs for mitigating NO$_X$ and PN emissions from conventional vehicles.

The findings suggest that mitigating both NO$_X$ and PN emissions through after-treatment systems in conventional vehicles is by several orders of magnitude less costly than doing so by BEVs and PHEVs. The costs of mitigating NO$_X$ emissions of conventional light-duty and heavy-duty vehicles by catalysts ranges between 800-3,800 EUR/t NO$_X$ and are comparable to those of mitigating NO$_X$ emissions in the manufacturing industry (Figure 23). For PN, mitigation costs range between 7-88 EUR/10^{17} particles (Figure 24).

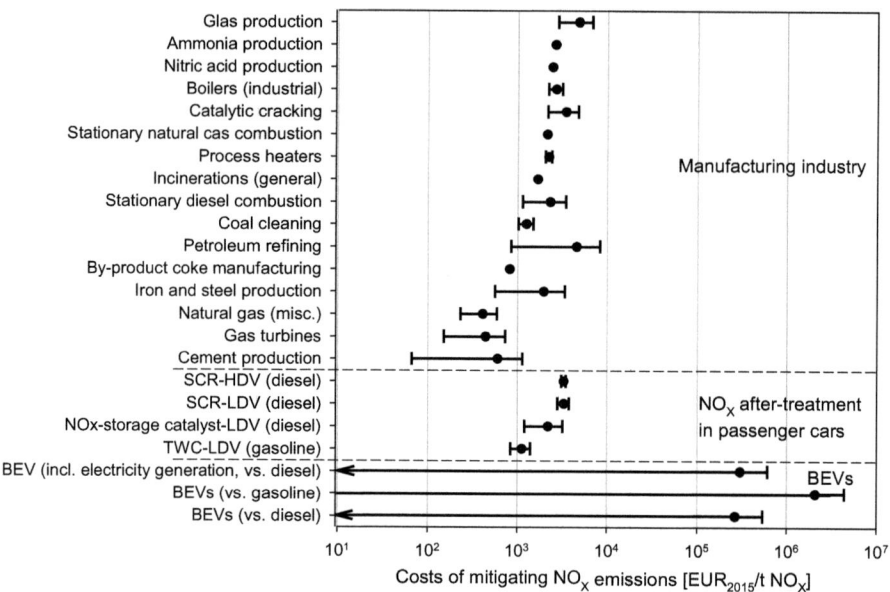

Figure 23: Costs of mitigating NOx emissions of conventional vehicles and in the manufacturing industry; dots depict the midpoint of cost ranges; costs of BEVs in 2016 span a wide range and may assume negative values, depicted here by arrows; SCR – selective catalytic reduction; LDV – light-duty vehicle; HDV – heavy-duty vehicles; TWC – three-way catalyst; only the three scenarios with the lowest mitigation costs are considered, therefore PHEVs are not shown, as their NOx mitigation is lower (Sources: see Table 20)

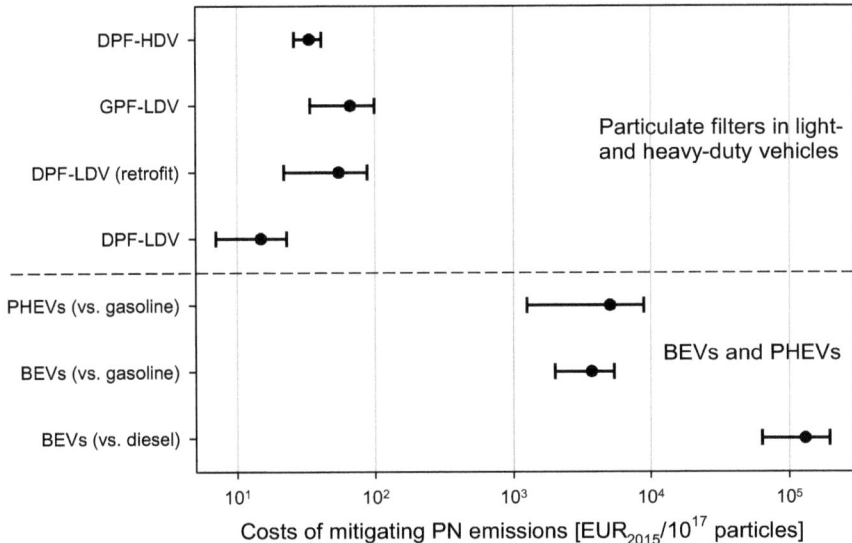

Figure 24: Costs of mitigating PN emissions of conventional vehicles; data for the manufacturing industry are not available; dots and error bars depict (i) the midpoint of cost ranges for particulate filters and (ii) the median and 25% of the interquartile range of cost values for BEVs and PHEVs for the year 2016; DPF – diesel particulate filter, GPF – gasoline particulate filter; LDV – light-duty vehicle; HDV – heavy-duty vehicles; PHEVs (vs. diesel) are not considered here, as we couldn't find savings of diesel-propelled PHEVs over diesel-propelled CVs (Sources: see Table 21)

These observations suggest that the further development of after-treatment systems for conventional vehicles with a focus on increasing efficacy and durability has economical merits and should be supported, e.g., by more stringent emissions standards. The high costs of mitigating air pollutants through BEVs and PHEVs suggest that supporting of these vehicles might be justified in view of decreasing the oil-dependency, CO_2 emissions, and noise pollution of road transport. However, low and near-zero pollutant emissions at the tailpipe might be, however, achieved in a less costly manner through advances in the after-treatment technologies in conventional vehicles.

The analysis presented here does not account for exposure to air pollutants. As humans tend to be more directly exposed to vehicle emissions than to emissions from industry and the power sector, pollution exposure should be accounted for when benchmarking the costs of mitigating emissions in the various economic sectors.

Table 20: Data sources and assumptions used for estimating the costs of mitigating NO$_x$ emissions of road vehicles and the manufacturing industry; data represent first-order estimates of typical costs that may not capture each specific case

Emissions source - Technology[d]	Costs	Lifetime	Efficacy	Mitigation costs
Light-duty vehicles – Three-way catalyst (including sensors and control software)	230-380 EUR (estimate based on Possada et al., 2012)[a,b]	240,000 km (based on Tier 3 durability requirements in the USA; Delphi, 2015)	95% NO$_x$ removal (based on Mooney (2007)	840-1,400 EUR/t NO$_x$
Light-duty vehicles – Lean NO$_x$-storage catalyst	250-680 EUR (estimate based on Possada et al., 2012)[a,b]	240,000 km based on Tier 3 durability requirements in the US (Delphi, 2015)	NO$_x$ reduction from 1.20 g/km to 0.30 g/km = 0,90 g/km (first order estimate of the authors)	1,200-3,200 EUR/t NO$_x$
Light-duty vehicles – Selective catalytic reduction	System costs: 330-500 EUR estimate based on Possada et al., 2012)[a,b]; Urea costs: 1.65 EUR/1000 km[c]	240,000 km based on Tier 3 durability requirments in the US (Delphi, 2015)	NO$_x$ reduction from 1.20 g/km to 0.12 g/km = 1,08 g/km (estimate of the authors)	2,800-3,800 EUR/t NO$_x$
Heavy-duty vehicles – Selective catalytic reduction	1,830-2,510 EUR (estimate of manufacturing costs for a vanadium-based SCR system based on Possada, et al., 2016)[a], urea costs: 0,88 EUR/100 km[c]	500,000 km based on Euro VI durability requirements (Delphi, 2017)	NO$_x$ reduction to 0.21 g/km (ICCT, 2016); NO$_x$ reduction efficiency of 95% (first order estimate of the authors)	3,100-3,500 EUR/t NO$_x$
Cement production – Various technologies (bio-solid injection, selective non-catalytic reduction, mid-kiln firing)	-	-	-	66-1,140 EUR/t NO$_x$[f] (EPA; 2015)
Gas turbines –	-	-	-	150-730 EUR/t NO$_x$[f]

Emissions source - Technology[d]	Costs	Lifetime	Efficacy	Mitigation costs
low NOx burner				(EPA; 2015)
Natural gas – pipeline compressors and stationary combustion (miscellaneous technologies including non-selective catalytic reduction, low-emission combustion, ignition retardation, adjusted air-to-fuel ratio)	-	-	-	230-590 EUR/t NOx[f] (EPA; 2015)
Iron and steel production – Low-NOx burning combined with flue gas recirculation, selective catalytic reduction	-	-	-	560-3,370 EUR/t NOx[f] (EPA; 2015)
By-product coke manufacturing – selective non-catalytic reduction	-	-	-	820 EUR/t NOx[f] (EPA; 2015)
Petroleum refining (incl. gas-fired processes) – selective catalytic reduction	-	-	-	850-8,310 EUR/t NOx[f] (EPA; 2015)
Coal cleaning – thermal drying and low NOx burning	-	-	-	1,020-1,490 EUR/t NOx[f] (EPA; 2015)
Stationary diesel and dual-fuel combustion (incl. for electricity generation) – ignition retardation, selective catalytic reduction	-	-	-	1,140-3,470 EUR/t NOx[f] (EPA; 2015)
Incinerators –	-	-	-	1,670 EUR/t NOx[f]

Emissions source - Technology[d]	Costs	Lifetime	Efficacy	Mitigation costs
selective non-catalytic reduction				(EPA; 2015)
Process heaters (gas fired and others) – ultra-low NO$_X$ burning, selective catalytic reduction	-	-	-	2,030-2,400 EUR/t NO$_X$[f] (EPA; 2015)
Natural gas stationary combustion for electricity generation – adjusted air to fuel ration and retarded ignition	-	-	-	2,130 EUR/t NO$_X$[f] (EPA; 2015)
Catalytic cracking, process heaters, coke ovens – flue gas recirculation, low-NO$_X$ burning	-	-	-	2,150-4,730 EUR/t NO$_X$[f] (EPA; 2015)
Boilers (Industrially-commercially-institutionally used, incl. coal and residual oil boilers) - low-NO$_X$ burning combined with flue gas recirculation, selective catalytic and non-catalytic reduction	-	-	-	2,190-3,140 EUR/t NO$_X$[f] (EPA; 2015)
Nitric acid production – non-selective catalytic reduction	-	-	-	2,430 EUR/t NO$_X$[f] (EPA; 2015)
Ammonia production – selective catalytic reduction	-	-	-	2,630 EUR/t NO$_X$[f] (EPA; 2015)
Glass manufacturing – OXY firing	-	-	-	2,820-6,800 EUR/t NO$_X$[f] (EPA; 2015)

Emissions source - Technology[d]	Costs	Lifetime	Efficacy	Mitigation costs
Miscellaneous industrial processes – flue gas recirculation, low-NO_X burning	-	-	-	3,660 EUR/t NO_X[f] (EPA; 2015)
Taconite ore processing – selective catalytic reduction	-	-	-	5,860 EUR/t NO_X[f] (EPA; 2015)

[a] We uniformly assume an exchange rate of 1.10 USD/EUR.
[b] We assume for the lower margin a cost reduction of 30% between 2012 and 2017.
[c] We assume here a consumption and price of urea solution of 1.5 l per 1,000 km and 1.10 EUR/l, respectively.
[d] In the case of after-treatment technologies for vehicle emissions, we assume technology levels necessary to comply with the Euro 6 emissions limits for light-duty vehicles. Considering production costs only.
[e] Rough estimate based on costs given by EPA (2015) in USD_{2011}; assuming an exchange rate of 1.10 USD/EUR.
[f] We assume here a consumption and price of urea solution of 1.6 l per 100 km and 0.55 EUR/l, respectively.

Table 21: Data sources and assumptions used for estimating the costs of mitigating PN emissions of road vehicles and the manufacturing industry; data represent first-order estimates of typical costs that may not capture each specific case

Emissions source – Technology[c]	Costs	Lifetime	Efficacy	Emission mitigation costs
Light-duty vehicles - Diesel particulate filter (including installation by car manufacturer)	150-350 EUR (estimate based on Possada Saches, 2012)[a,b]	200,000 km (based on FG, 2017; Giechaskiel, 2017	95% PN removal efficacy (Giechaskiel, 2017) to a level of $6*10^{11}$ particles/km (based on Giechaskiel et al., 2015)	7-23 EUR/10^{17} particles
Light-duty vehicles – Diesel particulate filter (retrofit by vehicle user)	500-2000 EUR (own estimate based on DHZ, 2016; FG, 2017)	200,000 km (based on FG, 2017; Giechaskiel, 2017	95% PN removal efficacy (Giechaskiel, 2017) to a level of $6*10^{11}$ particles/km (based on Giechaskiel et al., 2015)	22-88 EUR/10^{17} particles

Emissions source – Technology[e]	Costs	Lifetime	Efficacy	Emission mitigation costs
Light-duty vehicles – Gasoline particulate filter	67-158 EUR (Minjares and Posada Sanchez, 2011)[a,b,c]	200,000 km, assumption based on DPFs	70% PN removal (Giechaskiel, 2017); engine out emissions of $1,4*10^{12}$ particles/km decreased to $5-6*10^{11}$ particles/km (Bischof et al., 2012)	34-79 EUR/10^{17} particles
Heavy-duty vehicles – Diesel particulate filter	980-1,560 EUR (estimate of manufacturing costs for catalyzed DPFs based on Posada, et al., 2016)[d,e]	500,000 km based on Euro VI durability requirements (Delphi, 2017)	95% PN removal efficacy (Giechaskiel, 2017) to a level of $4*10^{11}$ particles/km (based on Giechaskiel et al., 2016 and an energy use of 1 kWh/km)	26-41 EUR/10^{17} particles

[a] We uniformly assume an exchange rate of 1.10 USD/EUR.
[b] We assume for the lower margin a cost reduction of 30% between 2012 and 2017.
[c] In the case of after-treatment technologies for vehicle emissions, we assume technology levels necessary to comply with the Euro 6 emissions limits for light-duty vehicles.
[d] Considering production costs only.
[e] Rough estimate based on costs given by EPA (2015) in USD_{2011}; assuming an exchange rate of 1.10 USD/EUR.